Rainwater Harvesting: Quantity, Quality, Economics and State Regulations

Special Issue Editor
Ataur Rahman

MDPI • Basel • Beijing • Wuhan • Barcelona • Belgrade

MDPI

Special Issue Editor
Ataur Rahman
University of Western Sydney
Australia

Editorial Office
MDPI AG
St. Alban-Anlage 66
Basel, Switzerland

This edition is a reprint of the Special Issue published online in the open access journal *Water* (ISSN 2073-4441) in 2015–2017 (available at: http://www.mdpi.com/journal/water/special issues/rainwater-harvest).

For citation purposes, cite each article independently as indicated on the article page online and as indicated below:

Lastname, F.M.; Lastname, F.M. Article title. *Journal Name*. **Year**. *Article number*, page range.

First Edition 2018

ISBN 978-3-03842-714-8 (Pbk)
ISBN 978-3-03842-713-1 (Pbk)

Table of Contents

About the Special Issue Editor v

Preface to "Rainwater Harvesting: Quantity, Quality, Economics and State Regulations" . . . vii

Ataur Rahman
Recent Advances in Modelling and Implementation of Rainwater Harvesting Systems
towards Sustainable Development
doi: 10.3390/w9120959 1

Liliana Lizrraga-Mendiola, Gabriela Vzquez-Rodrguez, Alberto Blanco-Pin, Yamile Rangel-
Martnez and Mara Gonzlez-Sandoval
Estimating the Rainwater Potential per Household in an Urban Area: Case Study in
Central Mexico
doi: 10.3390/w7094622 8

Chien-Lin Huang, Nien-Sheng Hsu, Chih-Chiang Wei and Wei-Jiun Luo
Optimal Spatial Design of Capacity and Quantity of Rainwater Harvesting Systems for
Urban Flood Mitigation
doi: 10.3390/w7095173 22

Yie-Ru Chiu, Yao-Lung Tsai and Yun-Chih Chiang
Designing Rainwater Harvesting Systems Cost-Effectively in a Urban Water-Energy Saving
Scheme by Using a GIS-Simulation Based Design System
doi: 10.3390/w7116285 48

Xiao Liang and Meine Pieter van Dijk
Identification of Decisive Factors Determining the Continued Use of Rainwater Harvesting
Systems for Agriculture Irrigation in Beijing
doi: 10.3390/w8010007 61

Lorena Liuzzo, Vincenza Notaro and Gabriele Freni
A Reliability Analysis of a Rainfall Harvesting System in Southern Italy
doi: 10.3390/w8010018 72

Peter Melville-Shreeve, Sarah Ward and David Butler
Rainwater Harvesting Typologies for UK Houses: A Multi Criteria Analysis of
System Configurations
doi: 10.3390/w8040129 92

Gerardo Smano-Romero, Marina Mautner, Alma Chvez-Meja and Blanca Jimnez-Cisneros
Assessing Marginalized Communities in Mexico for Implementation of Rainwater
Catchment Systems
doi: 10.3390/w8040140 110

Ammar Adham, Michel Riksen, Mohamed Ouessar and Coen J. Ritsema
A Methodology to Assess and Evaluate Rainwater Harvesting Techniques in (Semi-)
Arid Regions
doi: 10.3390/w8050198 124

Robert O. Ojwang, Jrg Dietrich, Prajna Kasargodu Anebagilu, Matthias Beyer and Franz
Rottensteiner
Rooftop Rainwater Harvesting for Mombasa: Scenario Development with Image
Classification and Water Resources Simulation
doi: 10.3390/w9050359 147

About the Special Issue Editor

Ataur Rahman is an Associate Professor of Hydrology in Western Sydney University, Australia. His research interest includes statistical hydrology, rainwater harvesting and flood hydrology. He has authored over 350 publications that include over 100 journal articles. He is serving in the editorial board of Water and Journal of Hydrologic Engineering. He received the G N Alexander Medal from Engineers Australia for his research in statistical hydrology.

Preface to "Rainwater Harvesting: Quantity, Quality, Economics and State Regulations"

Rainwater harvesting is a centuries old water supply technology and plays a major role to meet ever increasing water demand and cope with the climate change and variability. There has been a renewed interest on rainwater harvesting as a means of sustainable water resources management tool. The research on rainwater harvesting is getting broader covering aspects such as water savings and conservation, stormwater management, urban and rural agriculture, economic analysis and environmental issues. This special issue on rainwater harvesting is aimed at covering some of the emerging issues on rainwater harvesting.

The application of rainwater harvesting covers a broad range of geographical areas, uses and sustainability issues. In remote regions, rainwater harvesting is often used as a principal water supply tool for drinking, agriculture and sanitation purposes. In urban areas, rainwater harvesting is generally used as an alternative water supply means for the non-potable purposes and for control of stormwater. Rainwater harvesting is also used as a water source for small scale agricultural needs in both urban and rural areas. Rainwater harvesting is a primary water source in many rural areas and on islands. Wider implementation of rainwater harvesting system can delay the construction of new water supply infrastructures such as dam and pipeline. Rainwater harvesting enhances water availability for domestic and agricultural needs in semi-arid regions. In areas of increasing water scarcity, rainwater harvesting system can provide a more resilient and cost-efficient means of enhancing water security than complex public water supply system.

The effectiveness of rainwater harvesting system in water savings and conservations has been demonstrated across the globe covering a wide range of climatic conditions and applications. The quality of harvested rainwater largely depends on the surrounding environment, the tank material and maintenance of the rainwater harvesting system. Rainwater harvested from the roof catchments may contain heavy metals and nutrients. Use of adequately designed first flush device and regular maintenance of the rainwater harvesting system can significantly improve the harvested water quality.

The modelling of rainwater harvesting system seeks to match the rainwater availability with the projected water demand. Rainwater harvesting system is also analyzed as stormwater management component. Incorporation of environmental objectives and impact of climate change consideration into the design of rainwater harvesting system can significantly affect the determination of appropriate tank size.

The economic analysis of rainwater harvesting system needs to consider the cost implications of a whole range of issues such as quantity of water saved, water price, interest rate, environmental benefits, productive use and saved time for fetching water, the cost of alternative water supplies and maintenance of rainwater harvesting system.

To provide an update of some of the emerging issues on rainwater harvesting as discussed above, this special issue presents nine articles covering modelling, regionalization, uncertainty analysis, water-energy nexus, sustainability and urban flood mitigation. I believe that this special issue will be useful to researchers and policymakers on rainwater harvesting to be familiar with the current research and also to formulate future research tasks.

I would like to thank all the 33 authors for their contributions to this special issue, the reviewers for devoting their time and efforts to review the manuscripts and Water Editorial team for their great support during the review of the submitted manuscripts.

Ataur Rahman
Special Issue Editor

water

MDPI

Editorial

Recent Advances in Modelling and Implementation of Rainwater Harvesting Systems towards Sustainable Development

Ataur Rahman

School of Computing, Engineering and Mathematics, Western Sydney University, Locked Bag 1797, Penrith, NSW 2751, Australia; a.rahman@westernsydney.edu.au; Tel.: +61-2-4736-0145

Received: 9 October 2017; Accepted: 6 December 2017; Published: 8 December 2017

Abstract: Rainwater harvesting (RWH) is perhaps the most ancient practice to meet water supply needs. It has received renewed attention since the 1970s as a productive water source, water savings and conservation means, and sustainable development tool. In RWH, it is important to know how much water can be harvested at a given location from a given catchment size, whether the harvested water meets the intended water quality, whether the RWH system is economically viable and whether the state regulations favor the RWH. Furthermore, the selected RWH system should be suitable to local rainfall and field conditions, downstream impacts, and socio-economic and cultural characteristics. In this regard, this paper provides an overview of the special issue on "Rainwater Harvesting: Quantity, Quality, Economics and State Regulations". The selected papers cover a wide range of issues that are relevant to RWH such as regionalization of design curves, use of spatial technology, urban agriculture, arid-region water supply, multi criteria analysis and application of artificial neural networks.

Keywords: rainwater harvesting; water quality; water conservation; rainwater tanks; life cycle cost analysis; multi criteria analysis; urban flooding

---

1. Introduction

Rainwater harvesting (RWH) is a centuries old water supply technology and plays a major role to meet ever increasing water demand and cope with the climate change and variability. RWH is defined as a method of inducing, collecting, storing, and conserving local surface runoff for subsequent use. The RWH system collects rainwater from impervious surfaces (e.g., rooftops, terraces, courtyards and road surfaces) or natural land surface and stores water in a storage system such as tanks, cisterns and subsurface dams for both indoor and outdoor use [1,2]. In remote regions, RWH contributes towards meeting one of the targets of Sustainable Development Goals (ensuring availability and sustainable management of water and sanitation for all). In urban areas, the rainwater harvesting (RWH) system is generally used as an alternative water supply means for the non-potable purposes (e.g., toilet flushing, laundry, irrigation and car washing), and for control of stormwater [3,4]. The RWH system is also used as a water source for small scale agricultural needs in both urban and rural areas. RWH is a primary water source in many rural areas and on islands [4]. Wider implementation of the RWH system can delay the construction of new water supply infrastructures such as dams and pipelines. RWH enhances water availability for domestic and agricultural needs in semi-arid regions [2]. In areas of increasing water scarcity, the RWH system can provide a more resilient and cost-efficient means of enhancing water security than the complex public water supply system [5].

The evolution of the RWH system has been reviewed by a number of authors, e.g., [1,6–9]. The effectiveness of the RWH system in water savings and conservation has been demonstrated across the globe covering a wide range of climatic conditions and applications, e.g., in Australia [10], in Germany [11], in the USA [6], in Brazil [12], in the UK [13], in Italy [14], in China [15], in South Korea [16] and in West Asia and North Africa [9,17–20]. The quality of harvested rainwater largely depends on the surrounding environment, the tank material and maintenance of the RWH system. Rainwater harvested from the roof catchments may contain heavy metals and nutrients [21,22]. Use of adequately designed first flush devices and regular maintenance of the RWH system (such as washing of roof surfaces, gutters, tanks and first flush devices, inspecting for points of entry for mosquitoes and vermin and removing overhanging trees from the rooftop) can significantly improve the harvested water quality [23,24].

The modelling of the RWH system seeks to match the rainwater availability with the projected water demand [4,25]. This is generally undertaken by continuous simulation of the inflow and outflow [26–28] or using empirical relationships [4,29,30] or stochastic analysis [6,31] or a web-based tool that integrates geo-referenced rainfall patterns [32]. RWH system is also analyzed as a stormwater management component [33–35]. Incorporation of environmental objectives (e.g., greenhouse gas emission and materials used in construction of RWH system) into the design of the RWH system can significantly affect the determination of appropriate tank size [36]. Similarly, the impact of climate change consideration can also affect the tank size [37].

The economic analysis of RWH system needs to consider the cost implications of a whole range of issues such as quantity of water saved, water price, interest rate, environmental benefits, productive use and saved time for fetching water (which can be used for other productive uses, the cost of alternative water supplies and maintenance of the RWH system [7,9]. Previous studies on economic analysis of the RWH system have often demonstrated conflicting results depending on the issues considered in the analysis as explained in Campisano et al. [1] and Amos et al. [7]. Some researchers have shown that the RWH system is not financially viable in their considered scenarios [38–40], while other researchers have found that it is financially viable in specific cases [41–43]. These differences in conclusions can be attributed to the way maintenance and operational costs were incorporated into the analysis (e.g., energy cost to run the pump, maintenance of the RWH system and tank life), availability and cost of other sources of water and consideration of multiple benefits offered by the RWH system (e.g., productive water use [9] and environmental benefits).

The research on RWH is getting broader covering aspects such as water savings and conservation, stormwater management, urban agriculture, economic analysis and environmental issues. In this regard, this special issue contains nine articles covering modelling, regionalization, uncertainty analysis, water-energy nexus, sustainability and urban flood mitigation, the contents of which are analyzed in the following section.

2. Summary of This Special Issue

Use of spatial information in the modelling and analysis of RWH system is becoming popular. To supplement water supply in Mombasa, Ojwang et al. [44] demonstrated that a combination of satellite image analysis and modelling could be used as effective tools for developing RWH policy. In this study, image classification techniques were used to detect roof areas with reasonable accuracy. They also considered future population growth, improved living standards and future climate scenarios in the analysis. They noted that the RWH system could provide 2.3 to 23 million cubic meters of water, which, however, was not enough to meet the full water demand in the study area. This study will be a useful reference to RWH system design in future based on satellite image analysis.

The RWH system can offer significant water savings and conservation benefits even in the arid regions. For the Oum Zessar watershed in southeastern Tunisia, which is a semi-arid region, Adham et al. [45] developed an RWH system design methodology that integrated engineering, biophysical and socio-economic criteria using the Analytical Hierarchy Process (AHP) aided by the Geographic Information System (GIS). In this method, establishment of the scores/weighting for selected criteria was dependent on expert opinion. This integrated method was successfully applied to the study area, which provided suitability scores to the candidate RWH sites. The developed methodology was found to be highly flexible and easy to adapt to other regions. This methodology will be useful to designers and decision makers across different countries for enhancing the performance of existing and new RWH sites.

Regionalization of RWH system parameters assists the designers and planners in implementing the RWH system without the need of formal data analysis, which is often time-consuming. In this regard, Sámano-Romero et al. [46] used regional water access and precipitation data in Mexico to identify municipalities that would most benefit from the installation of domestic RWH systems. The developed method considered monthly rainfall data, number of occupants per household, water demand and run-off coefficient to calculate catchment area and tank size needed for a single dwelling. A curve was then developed to estimate catchment area based on annual rainfall for the selected municipalities that resulted in an average catchment area of 113.3 m^2 for a water demand of 100 L/capita/day. This study demonstrated that regional approximation could assist in the national implementation of the RWH system.

The RWH system offers many other benefits than savings of mains water. In this regard, Melville-Shreeve et al. [47] adopted a quantitative multi criteria analysis to assess the RWH system under a range of emerging criteria in the UK. They noted that traditional design approaches of the RWH system adopt whole life cost assessments that aim for financial savings associated with the provision of an alternative water supply, which disregard broader benefits of RWH system such as stormwater management issue. They proposed a number of RWH system configurations that would outperform traditional RWH system in relation to benefits and cost. The outcomes of this study will result in cost-effective implementation of RWH system in the UK. The approach can be adapted to other countries to demonstrate the wider benefit of RWH system.

In another RWH system regionalization study, Liuzzo et al. [48] assessed the reliability of using RWH system to provide water for toilet flushing and garden irrigation for a typical single-family home scenario in Sicily, Italy. Data on water consumption was collected and a daily water balance model was developed, and the model performance was evaluated using rainfall data over 100 stations in Sicily. Based on regional analysis, annual reliability curves for the RWH system as a function of mean annual rainfall was developed, which would be used by the designers and planners in the region. An uncertainty analysis was undertaken to evaluate the accuracy of the developed regional design curves. The benefit–cost analysis revealed that the implementation of an RWH system in Sicily could provide environmental and economic advantages over traditional water supply methods.

Irrigation by the RWH system in urban areas is getting renewal attention. On this issue, Liang et al. [49] evaluated the role of non-technological factors in RWH for agriculture irrigation in Beijing. In this study, 10 non-technological and technological impact factors were chosen and thereafter, based on an artificial data mining method and rough set analysis, the decisive factors were found. The most important finding of this study was that two non-technological factors, "doubts about rainwater quality" and "the availability of groundwater" largely defined the success of the RWH system in Beijing. They suggested that it is important to enhance public confidence and to motivate users on utilizing rainwater for agriculture irrigation to make the RWH system sustainable. This study highlights that non-technological factors such as public perception and motivation are important along with technological factors such as reliability and financial benefits of the RWH system.

The RWH system can contribute positively towards the water–energy nexus. In this regard, Chiu et al. [50] investigated the water–energy issue in relation to the RWH system in Taiwan. They presented a geographic information systems (GIS)-simulation-based design system to investigate whether the RWH system can be cost-effectively designed as an innovative water-energy conservation scheme on a regional level. They integrated a rainfall database, water balance model, spatial technologies, energy-saving investigation, and economic feasibility analysis for eight communities in Taipei, Taiwan. They exploited the temporal and spatial variations in rainfall to enhance the evaluation of the RWH system. The interesting finding of this study was that the RWH system became feasible based on the optimal design when both water and energy-savings were considered. They found that

RWH could achieve 21.6% domestic water-use savings and 138.6 (kWh/year-family) energy-savings. The findings of this study will assist RWH research in the water–energy nexus, which could also include urban agriculture supported by the RWH system to contribute towards the water–energy–food nexus.

The RWH system can be used to mitigate urban flooding, which is a relatively new area of RWH research. In this regard, Huang et al. [51] integrated the RWH system with the popular stormwater runoff management model (SWMM) in Zhong-He District, Taiwan. They adopted fuzzy C-means clustering to form similar subregions based on urban roof, land use and drainage systems. Based on statistical quartiles analysis for rooftop area and rainfall frequency analysis, they simulated the corresponding reduced flooding circumstances. They also applied a backpropagation neural network for developing a water level simulation model of urban drainage systems to substitute for SWMM, and a tabu search-based algorithm was adopted with the embedded backpropagation neural network based SWMM to optimize the planning solution. They found that the optimized spatial RWH system could reduce 72% of flood inundation losses based on the simulated flood events. The developed RWH modelling framework can be adapted to other cities having significant flooding problems.

In cities with limited rainfall and inadequate water supply, the RWH system can assist with solving water problems. In a study in Pachuca and Mineral de la Reforma, State of Hidalgo, Central Mexico by Lizárraga-Mendiola et al. [52] demonstrated that the harvestable rainwater from a roof area of 45 m^2 and 50 m^2 would be sufficient most of the year to meet toilet flushing and washing machine water demand; however, 100 m^2 and 200 m^2 roof area could provide enough water to meet other water demand too.

3. Conclusions

RWH contributes towards meeting one of the targets of the Sustainable Development Goals by serving as the principal water supply means for the remote and drought-prone regions and by saving a significant volume of mains water and offering substantial environmental benefits.

This special issue has covered a wide range of contemporary issues on RWH as summarized below. Development of design curves for the RWH system assists wider application of the RWH system in a region without the need of at-site data analysis as demonstrated by Sámano-Romero et al. [46] and Liuzzo et al. [48]. The feasibility and benefits of implementing the RWH system is likely to be underestimated if only monetary benefit is considered [50]. Other benefits, such as greenhouse gas emission reduction, urban stormwater runoff mitigation, reducing water stress during peak hours, and decreased demand on current water and energy facilities, should not be neglected in evaluating the RWH system. Use of spatial technology will make RWH system modelling more effective in identifying roof areas and other impervious areas and flood-prone areas for urban flood mitigation purpose as highlighted in Ojwang et al. [44], Adham et al. [45], Chiu et al. [50] and Huang et al. [51]. The RWH system can be used effectively for urban flood mitigation, which is a major problem in many cities around the world. The RWH modelling framework developed by Huang et al. [51] is capable of selecting a flexible and practical spatial arrangement and capacity design approach for RWH to serve as an alternative means for urban flood mitigation. The RWH system can assist in solving water supply problems even in arid regions as noted by Lizárraga-Mendiola et al. [52] and Adham et al. [45]. A similar conclusion was reached by Hajani and Rahman [28]. The RWH system can contribute positively to the water–energy–food nexus in urban areas, which is a relatively new area in RWH research [49,50]. The use of new data analysis techniques such as quantitative multi criteria analysis [47] and artificial neural networks [51] can demonstrate enhanced viability of the RWH system.

Further research on RWH should focus on financial analysis covering multiple benefits, life cycle analysis incorporating energy use and greenhouse gas emission, productive water use such as boosting rural and urban agriculture, and institutional and socio-political support to improve acceptability of RWH.

Water **2017**, *9*, 959

Acknowledgments: The author of this paper and editor of this special issue would like to thank three anonymous reviewers for their constructive suggestions to improve this editorial, all authors for their notable contributions to this special issue, the reviewers for devoting their time and efforts to reviewing the manuscripts and the Water Editorial team for their great support during the review of the submitted manuscripts.

Conflicts of Interest: The author declares no conflict of interest.

References

1. Campisano, A.C.; Butler, D.; Ward, S.; Burns, M.J.; Friedler, E.; DeBusk, K.; Fisher-Jeffes, L.N.; Ghisi, E.; Rahman, A.; Furumai, H.; et al. Urban rainwater harvesting systems: Research, implementation and future perspectives. *Water Res.* **2017**, *115*, 195–209. [CrossRef] [PubMed]
2. Lasage, R.; Verburg, P.H. Evaluation of small scale water harvesting techniques for semi-arid environments. *J. Arid Environ.* **2015**, *118*, 48–57. [CrossRef]
3. Van der sterren, M.; Rahman, A.; Dennis, G.R. Implications to stormwater management as a result of lot scale rainwater tank systems: A case study in Western Sydney, Australia. *Water Sci. Technol.* **2012**, *65*, 1475–1482. [CrossRef] [PubMed]
4. Hanson, L.S.; Vogel, R.M. Generalized storage–reliability–yield relationships for rainwater harvesting systems. *Environ. Res. Lett.* **2014**, *9*, 075007. [CrossRef]
5. Batchelor, C.; Fonseca, C.; Smits, S. *Life-Cycle Costs of Rainwater Harvesting Systems*; IRC International Water and Sanitation Centre: The Hague, The Netherlands, 2011; Volume 46, p. 37.
6. Basinger, M.; Montalto, F.; Lall, U. A rainwater harvesting systems reliability model based on nonparametric stochastic rainfall generator. *J. Hydrol.* **2010**, *392*, 105–118. [CrossRef]
7. Amos, C.C.; Rahman, A.; Gathenya, J.M. Economic analysis and feasibility of rainwater harvesting systems in urban and peri-urban environments: A review of the global situation with a special focus on Australia and Kenya. *Water* **2016**, *8*, 149. [CrossRef]
8. Pandey, D.N.; Gupta, A.K.; Anderson, D.M. Rainwater harvesting as an adaptation to climate change. *Curr. Sci.* **2003**, *85*, 46–59.
9. Bouma, J.A.; Hegde, S.S.; Lasage, R. Assessing the returns to water harvesting: A meta-analysis. *Agric. Water Manag.* **2016**, *163*, 100–109. [CrossRef]
10. Rahman, A.; Keane, J.; Imteaz, M.A. Rainwater harvesting in Greater Sydney: Water savings, reliability and economic benefits. *Resour. Conserv. Recycl.* **2012**, *61*, 16–21. [CrossRef]
11. Schuetze, T. Rainwater harvesting and management—Policy and regulations in Germany. *Water Sci. Technol. Water Supply* **2013**, *13*, 376–385. [CrossRef]
12. Ghisi, E.; Bressan, D.L.; Martini, M. Rainwater tank capacity and potential for potable water savings by using rainwater in the residential sector of southeastern Brazil. *Build. Environ.* **2007**, *42*, 1654–1666. [CrossRef]
13. Ward, S.; Butler, S. Rainwater harvesting and social networks: Visualizing interactions for niche governance, resilience and sustainability. *Water* **2016**, *8*, 526. [CrossRef]
14. Campisano, A.; Gnecco, I.; Modica, C.; Palla, A. Designing domestic rainwater harvesting systems under different climatic regimes in Italy. *Water Sci. Technol.* **2013**, *67*, 2511–2518. [CrossRef] [PubMed]
15. Gould, J.; Zhu, Q.; Yuanhong, L. Using every last drop: Rainwater harvesting and utilization in Gansu Province, China. *Waterlines* **2014**, *33*, 107–119. [CrossRef]
16. Han, M.Y.; Mun, J.S. Operational data of the Star City rainwater harvesting systems and its role as a climate change adaptation and a social influence. *Water Sci. Technol.* **2011**, *63*, 2796–2801. [CrossRef] [PubMed]
17. Ziadat, F.; Bruggeman, A.; Oweis, T.; Haddad, N.; Mazahreh, S.; Strtawi, W.; Syuof, M. A participatory GIS approach for assessing land suitability for rainwater harvesting in an arid rangeland environment. *Arid Land Res. Manag.* **2012**, *26*, 297–311. [CrossRef]
18. Oweis, T.; Hachum, A. Water harvesting and supplemental irrigation for improved water productivity of dry farming systems in West Asia and North Africa. *Agric. Water Manag.* **2006**, *80*, 57–73. [CrossRef]
19. Bruins, H.J.; Evenari, M.; Nessler, U. Rainwater harvesting agriculture for food production in arid zones: The challenge of the African famine. *Appl. Geogr.* **1986**, *6*, 13–32. [CrossRef]
20. Shariti, E.; Unami, K.; Mohawesh, O.; Fujihara, M. Design and construction of a hydraulic structure for rainwater harvesting in arid environment. In Proceedings of the 36th IAHR World Congress, Delft, The Netherlands, 28 June–3 July 2015.

21. Van der Sterren, M.; Rahman, A.; Dennis, G. Quality and quantity monitoring of five rainwater tanks in Western Sydney, Australia. *J. Environ. Eng.* **2013**, *139*, 332–340. [CrossRef]
22. Hamdan, S.M. A literature based study of stormwater harvesting as a new water resource. *Water Sci. Technol.* **2009**, *60*, 1327–1339. [CrossRef] [PubMed]
23. Melidis, P.; Akratos, C.S.; Tsihrintzis, V.A.; Trikilidou, E. Characterization of rain and roof drainage water quality in Xanthi, Greece. *Environ. Monit. Assess.* **2007**, *127*, 15–27. [CrossRef] [PubMed]
24. Abdulla, F.A.; Al-Shareef, A.W. Roof rainwater harvesting systems for household water supply in Jordan. *Desalination* **2009**, *243*, 195–207. [CrossRef]
25. Hajani, E.; Rahman, A. Reliability and cost analysis of a rainwater harvesting systems in peri-urban regions of Greater Sydney, Australia. *Water* **2014**, *6*, 945–960. [CrossRef]
26. Ward, S.; Memon, F.A.; Butler, D. Performance of a large building rainwater harvesting systems. *Water Res.* **2012**, *46*, 5127–5134. [CrossRef] [PubMed]
27. Sample, D.J.; Liu, J. Optimizing rainwater harvesting systems for the dual purposes of water supply and runoff capture. *J. Clean. Prod.* **2014**, *75*, 174–194. [CrossRef]
28. Hajani, E.; Rahman, A. Rainwater utilization from roof catchments in arid regions: A case study for Australia. *J. Arid Environ.* **2014**, *111*, 35–41. [CrossRef]
29. Ghisi, E. Parameters influencing the sizing of rainwater tanks for use in houses. *Water Resour. Manag.* **2010**, *24*, 2381–2403. [CrossRef]
30. Eroksuz, E.; Rahman, A. Rainwater tanks in multi-unit buildings: A case study for three Australian cities. *Resour. Conserv. Recycl.* **2010**, *54*, 1449–1452. [CrossRef]
31. Unami, K.; Mohawesh, O.; Sharifi, E.; Takeuchi, J.; Fujihara, M. Stochastic modelling and control of rainwater harvesting systems for irrigation during dry spells. *J. Clean. Prod.* **2015**, *88*, 185–195. [CrossRef]
32. Fonseca, C.R.; Hidalgo, V.; Díaz-Delgado, C.; Vilchis-Francés, A.Y.; Gallego, I. Design of optimal tank size for rainwater harvesting systems through use of a web application and geo-referenced rainfall patterns. *J. Clean. Prod.* **2017**, *145*, 323–335. [CrossRef]
33. Van der sterren, M.; Rahman, A.; Shrestha, S.; Barker, G.; Ryan, G. An overview of on-site retention and detention policies for urban stormwater management in the greater Western Sydney region in Australia. *Water Int.* **2009**, *34*, 362–372. [CrossRef]
34. Campisano, A.; Modica, C. Appropriate resolution timescale to evaluate water saving and retention potential of rainwater harvesting for toilet flushing in single houses. *J. Hydroinform.* **2015**, *17*, 331–346. [CrossRef]
35. DeBusk, K.M.; Hunt, W.F.; Wright, J.D. Characterizing rainwater harvesting performance and demonstrating stormwater management benefits in the humid southeast USA. *J. Am. Water Resour. Assoc.* **2013**, *49*, 1398–1411. [CrossRef]
36. Morales-Pinzon, T.; Rieradevall, J.; Gasol, C.M.; Gabarrell, X. Modelling for economic cost and environmental analysis of rainwater harvesting systems. *J. Clean. Prod.* **2015**, *87*, 613–626. [CrossRef]
37. Haque, M.M.; Rahman, A.; Samali, B. Evaluation of climate change impacts on rainwater harvesting. *J. Clean. Prod.* **2016**, *137*, 60–69. [CrossRef]
38. Kumar, M.D. Roof water harvesting for domestic water security: Who gains and who loses? *Water Int.* **2004**, *29*, 43–53. [CrossRef]
39. Roebuck, R.M.; Oltean-Dumbrava, C.; Tait, S. Whole life cost performance of domestic rainwater harvesting systems in the United Kingdom. *Water Environ. J.* **2011**, *25*, 355–365. [CrossRef]
40. Rahman, A.; Dbais, J.; Imteaz, M. Sustainability of rainwater harvesting systems in multistory residential buildings. *Am. J. Eng. Appl. Sci.* **2010**, *3*, 889–898. [CrossRef]
41. Imteaz, M.A.; Shanableh, A.; Rahman, A.; Ahsan, A. Optimisation of rainwater tank design from large roofs: A case study in Melbourne, Australia. *Resour. Conserv. Recycl.* **2011**, *55*, 1022–1029. [CrossRef]
42. Domenech, L.; Saurí, D. A comparative appraisal of the use of rainwater harvesting in single and multi-family buildings of the Metropolitan Area of Barcelona (Spain): Social experience, drinking water savings and economic costs. *J. Clean. Prod.* **2011**, *19*, 598–608. [CrossRef]
43. Ghisi, E.; Schondermark, P.N. Investment feasibility analysis of rainwater use in residences. *Water Resour. Manag.* **2013**, *27*, 2555–2576. [CrossRef]
44. Ojwang, R.O.; Dietrich, J.; Anebagilu, P.K.; Beyer, M.; Rottensteiner, F. Rooftop rainwater harvesting for Mombasa: Scenario development with image classification and water resources simulation. *Water* **2017**, *9*, 359. [CrossRef]

45. Adham, A.; Riksen, M.; Ouessar, M.; Ritsema, C. A methodology to assess and evaluate rainwater harvesting techniques in (semi-) arid regions. *Water* **2016**, *8*, 198. [CrossRef]
46. Sámano-Romero, G.; Mautner, M.; Chávez-Mejía, A.; Jiménez-Cisneros, B. Assessing marginalized communities in Mexico for implementation of rainwater catchment systems. *Water* **2016**, *8*, 140. [CrossRef]
47. Melville-Shreeve, P.; Ward, S.; Butler, D. Rainwater harvesting typologies for UK houses: A multi criteria analysis of system configurations. *Water* **2016**, *8*, 29. [CrossRef]
48. Liuzzo, L.; Notaro, V.; Freni, G. A reliability analysis of a rainfall harvesting system in Southern Italy. *Water* **2016**, *8*, 18. [CrossRef]
49. Liang, X.; Dijk, M.P.V. Identification of decisive factors determining the continued use of rainwater harvesting systems for agriculture irrigation in Beijing. *Water* **2016**, *8*, 7. [CrossRef]
50. Chiu, Y.R.; Tsai, Y.L.; Chiang, Y.C. Designing rainwater harvesting systems cost-effectively in a urban water-energy saving scheme by using a GIS-simulation based design system. *Water* **2015**, *7*, 6285–6300. [CrossRef]
51. Huang, C.L.; Hsu, N.S.; Wei, C.C.; Luo, W.J. Optimal spatial design of capacity and quantity of rainwater harvesting systems for urban flood mitigation. *Water* **2015**, *7*, 5173–5202. [CrossRef]
52. Lizárraga-Mendiola, L.; Vázquez-Rodríguez, G.; Blanco-Piñón, A.; Rangel-Martínez, Y.; González-Sandoval, M. Estimating the rainwater potential per household in an urban area: Case study in Central Mexico. *Water* **2015**, *7*, 4622–4637. [CrossRef]

water

MDPI

Article

Estimating the Rainwater Potential per Household in an Urban Area: Case Study in Central Mexico

Liliana Lizárraga-Mendiola [1],*, Gabriela Vázquez-Rodríguez [2], Alberto Blanco-Piñón [3], Yamile Rangel-Martínez [1] and María González-Sandoval [1]

[1] Área Académica de Ingeniería, Universidad Autónoma del Estado de Hidalgo, Carr. Pachuca-Tulancingo km 4.5 Col. Carboneras, Mineral de la Reforma 42184, Mexico; yamilerangelm@gmail.com (Y.R.-M.); cuquisglezs@gmail.com (M.G.-S.)

[2] Área Académica de Química, Universidad Autónoma del Estado de Hidalgo, Carr. Pachuca-Tulancingo km 4.5 Col. Carboneras, Mineral de la Reforma 42184, Mexico; g.a.vazquezr@gmail.com

[3] Área Académica de Ciencias de la Tierra y Materiales, Universidad Autónoma del Estado de Hidalgo, Carr. Pachuca-Tulancingo km 4.5 Col. Carboneras, Mineral de la Reforma 42184, Mexico; blanco.abp@gmail.com

* Author to whom correspondence should be addressed; lililga.lm@gmail.com; Tel.: +52-771-717-2000 (ext. 4001).

Academic Editor: Ataur Rahman

Received: 24 June 2015; Accepted: 13 August 2015; Published: 27 August 2015

Abstract: In cities with problems of aridity and a shortage of drinking water supply, there is an urgent need to establish alternatives for an adequate water management program. This study proposes an estimation through which users can select a rainwater harvesting system for non-drinking water consumption. For the cities of Pachuca and Mineral de la Reforma, State of Hidalgo, Central Mexico, the historical record of rainfall analyzed covers a period of 33 years (1980–2013). We calculated the monthly volume of rainwater harvestable from roof areas (VR, m^3) with household roof areas (Hra) of 45 m^2, 50 m^2, 100 m^2 and 200 m^2. It is proposed to replace in each single house the flush toilets and washing machine with ecological devices with consumptions of 4.8 L/flush and 70 L/load, respectively. Furthermore, a maximum and a minimum consumption of eight and six flushes/day/person (flush toilets) and five and four loads/week (washing machine), respectively, are proposed. From these considerations, our estimations of the harvestable rainwater showed that households with Hra of 45 m^2 and 50 m^2 would depend on the water supply system of the public network during part of the year. On the other hand, households with Hra of 100 m^2 and 200 m^2 might be able to store enough water to meet other needs besides toilet flushing and laundry.

Keywords: domestic consumption; harvesting; Central Mexico; rainwater; roof area

1. Introduction

Population growth, urbanization and global climate change represent a very important pressure on urban water resources. These factors require that the water administrators consider immediately other options that counteract the water stress that the population is facing [1]. For this reason, it is becoming increasingly more recurrent to take advantage of rainwater in urban areas, mainly to meet consumption needs for which the use of drinking water is not imperative [2].

The African Development Bank [3] defines rainwater harvesting as "the collection of the runoff for productive use", particularly in areas where rainfall varies between 200 and 1000 mm; while for Sapkota *et al.* [4] and Liaw and Chang [5], it is the collection and use of rainwater for domestic purposes. Villarreal and Dixon [6] mentioned that, although in Sweden, only 0.5% of the available water is used, large amounts of rainwater are consumed at the household level in an area where the annual rainfall is as high as 508 mm. These authors studied a housing development made up of buildings, where the rainwater was collected from roof areas. They identified that the volume of rainwater harvested would help to save a significant amount of drinking water, especially if also ecological or water-saving devices replaced the regular flush toilets.

Khastagir and Jayasuriya [7] conducted a study in Melbourne, Southwestern Australia, where rainfall varies from 450 to 1050 mm per year and a drought of 12 consecutive years (1997–2009) has been experienced. These authors designed a methodology to determine the size of tanks of rainwater storage at the household level, considering that the distribution of rainfall can vary from one point to another of the city. Concerning their storage capacity, Imteaz *et al.* [8,9] and Rahman *et al.* [10] indicated that daily rainfall analysis is expected to produce more realistic rainwater tank sizing than using monthly rainfall data.

Through a probabilistic relationship, Su *et al.* [11] found that the efficiency of rainwater collecting systems depends on the temporal distribution of rainfall and the water demand. Jones and Hunt [12] designed a system for rainwater harvesting based on rainfall historical records in two cities of North Carolina, Southeastern United States. Their main interest was to determine an optimal balance between the roof surface, the size of the tank, as well as the water consumption.

In Mexico, the harvesting of rainwater would make a major contribution to reduce the water supply shortage that occurs in large areas of the country. In Guanajuato, Central Mexico, a project was conducted to harvest rainwater using the roof areas of the houses in a community with an average annual rainfall of 455.3 mm. Water storage tanks of a 2.5 m^3 capacity were installed in roofs of 74 m^2 [13]. In Mexico City and in rural areas of the country, there are hundreds of catchment systems already installed [14]. All of these study cases have had successful results at both the individual and local levels.

In 2002, Biswas [15] indicated that the federal government in Mexico had solved the problems of water shortage through the development of infrastructure intended to increase the water supply. However, the same author mentioned that federal institutions do not consider the social, cultural and economic conditions of one of the most populated regions of the country, *i.e.*, Mexico City. The reason is that these measures are often implemented when the problems are already critical or when the situation is unsustainable. Mazari-Hiriart and Mazari-Menzer [16] and Salazar-Adams and Pineda-Pablos [17] agreed on the impact that the increase in the population of Mexico has had on the natural water availability. In 1950, the person availability was 17,742 m^3 in this country, while in 2013, it decreased to 3982 m^3 [18]. At the current rate of population growth, in 2030, the water availability will be reduced to 3783 m^3 per person [18].

The cities of Pachuca and Mineral de la Reforma (Central Mexico) are located in the northern zone of the Cuautitlan-Pachuca aquifer, and their population shares the water pumped from it with the population of the metropolitan area of Mexico City. In accordance with Neri-Ramírez *et al.* [19], the recharge of this aquifer was 356.7 Hm^3/year during 2009. In agreement with these authors, the average annual abatement of the static level during the past 40 years has been 2.1 m/year, which has reduced its recharge ability by 45%. The main reason for this problem is the growth of urban sprawl and the rapid land use change, which have reduced the capacity of infiltration. Galindo-Castillo *et al.* [20] pointed out that the northern part of this aquifer presents the greatest risk of overexploitation. In addition, the preservation of the hydrological resource represents a major challenge to the authorities of these two cities, because this aquifer is one of their main sources of water supply.

Amaya-Ventura [21] carried out an analysis of the water management system in these cities and found several problems at the administration level, leading to frequent shortages of the water supply. Nowadays, the same problems are still present. For example, it was necessary to limit the service some hours per day due to poor planning in the system of distribution and supply, which has not been modernized. This type of limited service represents up to 40% of the capacity of the service distributed to the population.

Due to the foregoing, there is an urgent need to propose strategies helping to lessen the problem of water shortage that the population of Pachuca and Mineral de la Reforma are currently facing.

9

To this end, the objective of this paper is to propose an estimation through which users can select a rainwater harvesting system for non-drinking water consumption. Furthermore, depending on their catchment area, we calculated the amount of non-drinking water required (*i.e.*, the water consumed by flush toilets or washing machines) that can be replaced with rainwater.

2. Description of the Study Area

2.1. Location

Pachuca and Mineral de la Reforma are located in the south-central region of the State of Hidalgo, approximately 80 km to the north of the Mexico City metropolitan area. The city of Pachuca is placed between the coordinates 20°07′21″ north latitude and 98°44′09″ west longitude. The city of Mineral de la Reforma is located between the coordinates 20°08′08″ north latitude and 98°40′19″ west longitude (Figure 1). Due to its proximity, both cities share hydraulic infrastructure and the same local water supply system.

Figure 1. Location of the study area.

2.2. Hydrology

The area of Pachuca-Mineral de la Reforma is located within the Pánuco Basin, regionally within the basin of Mexico, and internally in the sub-basin Río de las Avenidas [22]. The average natural availability of water (year 2013) was 152 m³/person. This availability corresponds to "water stress" [13]. Runoffs are predominantly of dendritic morphology and have their origin in the peaks of the Sierra of Pachuca. The most important surface runoff is known as Río de las Avenidas, which passes through the center of the city of Pachuca in the NE-SW direction. This river behaves as an intermittent stream during most of the year, although after extraordinary rainfalls, it becomes an important drain for the city.

2.3. Climate

In both locations, the average annual rainfall is 376.96 mm (1980–2013); the average temperature is 16 °C/year; and the rainy season occurs mainly from May to September [23].

2.4. Population and Drinking Water Service

The population densities of the cities of Pachuca and Mineral de la Reforma are 1371 and 1201.92 person/km², respectively; both cities have an average annual growth rate of 2.16% and 8.48%, respectively; and total population is 267,862 and 105,870, respectively [24,25].

A total of 134,053 households having a drinking water supply was estimated in the study area. Among these, only 42% has a continuous service (24 h a day); the remaining 58% has a limited water supply, which can vary from a few hours a day throughout the week to only three times per week [25].

3. Materials and Methods

3.1. Temporary Distribution of Rainwater

To study the potential for rainwater harvesting from roof areas, daily, monthly and annual average rainfall data for the years 1980–2013 were analyzed [23]. From this period, the rainiest year was selected to estimate monthly rainwater harvesting (volume of rainwater (*VR*)). In addition, the main patterns of rain and drought during the studied period were identified. The intra-annual rainfall variation was determined through the coefficient of variation of the monthly rainfall (*CVm*) according to Aladenola and Adeboye [26] (Equation (1)):

$$CVm = \frac{Sv}{Va} \tag{1}$$

where:

CVm is the coefficient of variation of the monthly rainfall;
Sv is the standard deviation of the monthly rainfall (mm);
Va is the mean of the monthly rainfall (mm).

Furthermore, a descriptive statistical analysis was applied to rainfall data to examine their central tendency (mean, asymmetry and variance), variability (standard deviation) and peakedness (kurtosis). We assumed the annual monthly maximum rainfall data as a normal distribution and considered a single-tailed test. This analysis was performed according to the methodology proposed by Ahammed *et al.* [27]. Standardization of data was performed in order to eliminate potential data redundancy and inconsistent dependencies in a historic record rainfall (1980–2013) based on Ahammed *et al.* [27].

3.2. Potential of Rainwater Harvesting and Water Demand per Household

The potential of monthly rainwater harvesting from rooftops (*VR*) at the household level was determined using the method proposed by Aladenola and Adeboye [27] (Equation (2)):

$$VR = \frac{R \times Hra \times Rc}{1000} \tag{?}$$

where:

VR is the monthly volume of rainwater harvested per household (m³);
R is the monthly rainfall depth (mm);
Hra is the household roof area (m²);
Rc is the runoff coefficient (without units) = 0.70. This value indicates a 30% loss.

For the design of the systems for rainwater harvesting and storage, United Nations Environment Programme (UNEP) [28] recommends considering the "first flush" by subtracting the first 0.50 mm of rainfall. Khastagir and Jayasuriya [7] and Su *et al.* [11] suggested subtracting the first 0.33 mm of the daily rainfall to improve of the quality of the water stored. For this study, the first 0.33 mm of rainfall were subtracted.

To determine the household roof area available for rainwater harvesting, we identified the main types of household prevailing in Mexico. The National Housing Federal Agency [29] classifies four types of household roof areas according to their socioeconomic level: (1) social (45 m²); (2) popular (50 m²); (3) middle residential (100 m²) and (4) residential (200 m²); see Table 1. Furthermore, it was taken into account that the roof material is usually concrete.

The average water consumption required in households (*Wnc*, m³/month) was calculated from Equation (3), according to Khastagir and Jayasuriya [7]:

$$Wnc = (Wcpc \times n)/1000 \tag{3}$$

where: *Wcpc* is the water consumption per person. According to the local water administration, this is 125 L/person/day [25]; *n* is the number of persons per household [29].

Table 1. Classification of the household type and average water consumption required (*Wnc*) per month [29]. *Hra* = household roof area, m²; *n* = number of person per household; *Wnc* = average water consumption required in households per month.

Hra (m²)	*n*	*Wnc* (m³/month)
45	3.7	14.067
50	4.1	15.588
100	4.5	17.109
200	5.1	19.390

In addition to the household roof area (*Hra*), the average number of person per household (*n*) and the amount of water required for consumption per person per day (*Wcpc*), the average daily consumption of flush toilets and the average weekly consumption of washing machines considered as water-saving or ecological were estimated according to National Institute of Ecology and Climatic Change [30]. Both types of ecological device were selected because they represent an important consumption of water in any household and do not require drinking water for their operation. Furthermore, this is a simple practice to implement among the population. For this analysis, two scenarios were considered: (1) the minimum (Avc$_{min}$) and (2) the maximum (Avc$_{max}$) number of times per week that the flush toilets and the washing machine are used (Table 2).

Table 2. Average consumption of ecological devices (Avc) [29]. Avc$_{max}$ = average maximum number of times; Avc$_{min}$ = average minimum number of times.

Ecological Device	Consumption in L	Avc$_{max}$	Avc$_{min}$
Flush toilet	4.8 L/flush	8 flushes/day/person	6 flushes/day/person
Washing machine	70 L/load	5 loads/week	4 loads/week

From these data, water consumption was assessed weekly and monthly (over twelve months of the year) considering the separate use of the flush toilet and the washing machine, as well as the total consumption of the two ecological devices combined (the sum of flush toilet and washing machine consumptions). The basic monthly balance (m³/month) was estimated using the Equation (4):

$$Wa = VR - Iv \tag{4}$$

where:

Wa is the available water (m³/month);
VR is the monthly volume of rainwater harvested from roof areas (m³);

12

Iv is the initial volume in storage that is equal to the monthly volume necessary for the flush toilet, washing machine and their combined use (m^3).

4. Results and Discussion

4.1. Rainfall Temporary Distribution

The bimodal rainfall pattern in the study area is shown in Figure 2. The highest variability occurs during the rainiest months (May–October), while months with the lowest variability are the driest (rainfall from 0 to 50 mm). Anaya-Garduño [31] mentioned that for every 100 mm of rainfall in catchment surfaces of 100 m^2, it is possible to collect up to 10 m^3. These observations indicate that the rainwater harvesting practice can be a good option to alleviate water supply deficiencies in the study area.

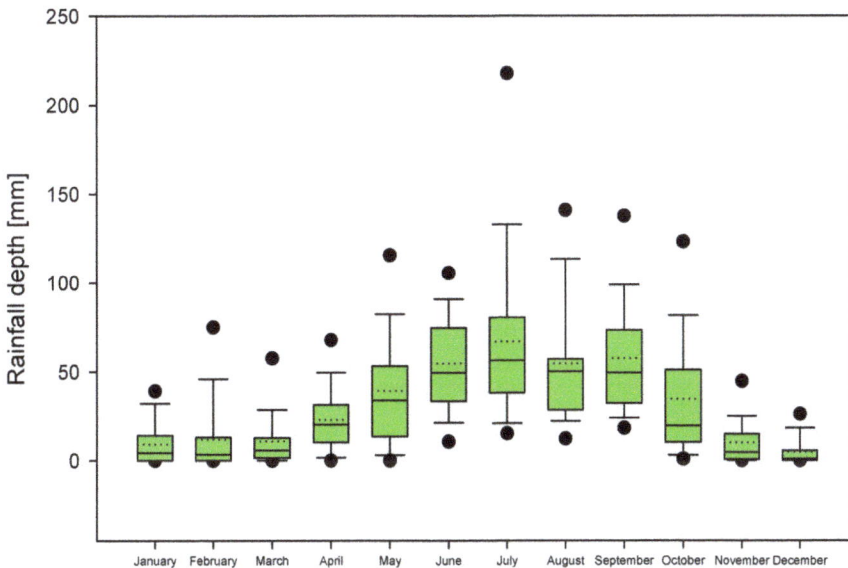

Figure 2. Monthly rainfall distribution. The dotted and continuous lines inside the boxes represent the mean and median values, respectively. The boundaries of the boxes represent the 25th and 75th percentiles. Error bars above and below the boxes indicate the 90th and 10th percentiles, respectively. The points represent the outline values measured for each month.

The behavior of rainfall through the historical standardized period (1980–2013) shows the following: the standard deviation and variance values are one, and the kurtosis value is 0.149400907. These values indicate that in comparison to the normal distribution, the central peak is flatter and wider (Table 3).

Table 3. Descriptive statistics of the standardized rainfall pattern in the study area.

Annual Average Rainfall (mm)	
Mean	1.56737×10^{-16}
Variance	1
Asymmetry	−0.030880758
Kurtosis	0.149400907
Standard deviation	1

The lowest average annual rainfall was recorded in 1982 (181.10 mm), whereas the highest average annual rainfall was recorded in 2010 (585.60 mm). The average rainfall for the historical period is 376.96 mm. The variability analysis through the historical period indicates that rainfall started to increase since 1997 to the present (Figure 3). If this tendency continues, it represents a good opportunity to implement an extensive rainwater harvesting system at the household level in order to save potable water through its substitution for some specific purposes (flush toilet, laundry, garden irrigation, among others). For this study, the rainiest year (2010) was selected to estimate monthly VR at the household level (Figure 4). The rainwater harvesting in areas with low precipitation (508 mm) has been demonstrated to have successful results by Villarreal and Dixon [6]. They identified that the volume of rainwater harvested from the roof area of a residential building would help to save a significant amount of drinking water, especially if also ecological or water-saving devices replaced the flush toilets. In other study, in Guanajuato, Central Mexico, United Nations Development Programme [13] developed a project in a small town with 455.3 mm of rainfall, and their results indicated that it is possible to store 2.5 m^3 at the household level.

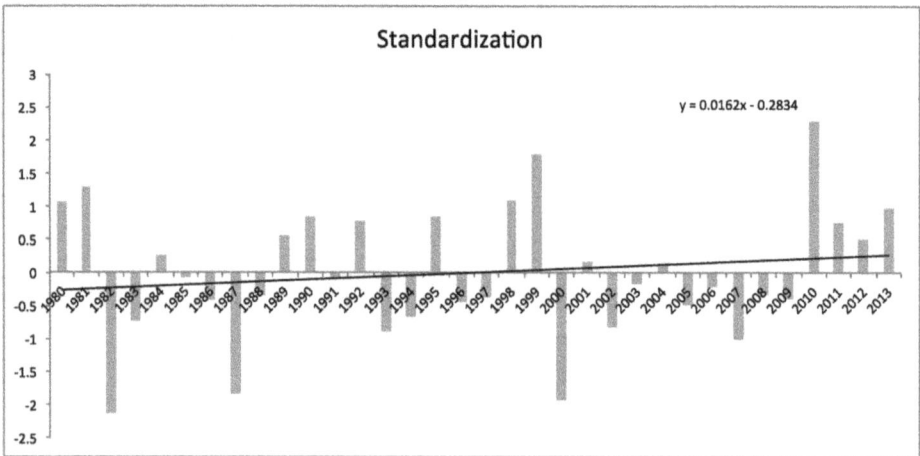

Figure 3. Standardized trend of the historical record of rainfall in mm based on Ahammed *et al.* [27]. Y axis: Frequency of rainfall events for the period 1980–2013.

After the first 0.33 mm were subtracted from the daily rainfall in this study, the annual monthly variation (year 2010) was analyzed (Figure 3). The results from the analysis show a non-normal behavior, which indicates that during nine months of the year, the rainfall varies from 0 mm to 50 mm; two months per year present between 50 mm to 100 mm of rainfall; and only one month receives between 200 mm to 250 mm (Figure 4).

The intra-annual variation was calculated (CVm) and ranged between 0.50 and 1.73. This difference shows that there is a high variability in the rainfall distribution [26]. This high variability could reduce rainwater harvesting potential during the driest months; however, if there is water availability during the rainiest months, water can be saved and used further.

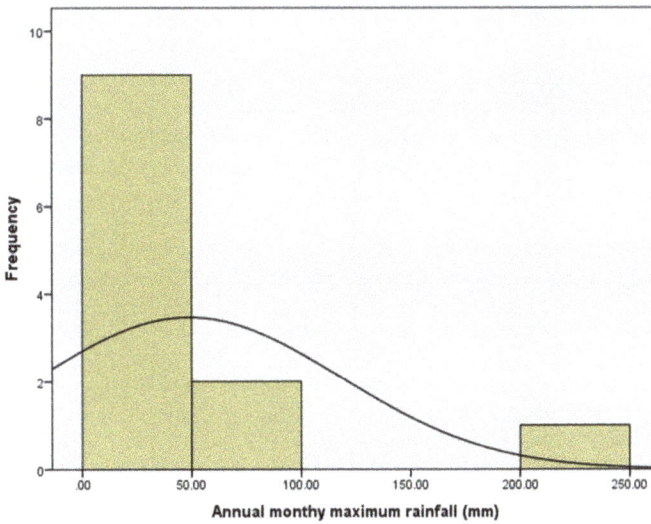

Figure 4. Histogram of annual monthly maximum rainfall data.

4.2. Potential of Rainwater Harvesting and Water Consumption per Household

The volume of rainwater harvestable from roof areas (VR, in m^3) that could be collected monthly according to the type of household roof area (Hra) is listed in Table 4.

Table 4. Monthly volume of rainwater harvestable (VR, in m^3) as a function of the household roof area (Hra = 45 m^2, 50 m^2, 100 m^2 and 200 m^2).

Month	45 m^2	50 m^2	100 m^2	200 m^2
January	1.010	1.122	2.244	4.489
February	2.349	2.610	5.219	10.439
March	1.908	2.120	4.239	8.479
April	2.150	2.389	4.778	9.557
May	4.050	4.500	8.999	17.999
June	4.009	4.454	8.908	17.817
July	7.540	8.378	16.755	33.511
August	4.626	5.140	10.280	20.561
September	6.664	7.405	14.809	29.619
October	5.401	6.001	12.002	24.005
November	2.238	2.487	4.974	9.949
December	0.720	0.800	1.600	3.201

Table 4 shows that even during the months with less rainfall, it is still possible to harvest rainwater. For instance, in the driest month (December), it is possible to store a limited water volume (less than 1 m^3), even in households with Hra of 45 m^2 and 50 m^2. However, this low volume of water harvested can be compensated by the water stored during the rainy season.

The volume of water required for flush toilets and washing machines, as well as for both uses combined (Table 5) was estimated considering the maximum and minimum consumption necessary for ecological devices (Avc, m^3/month) (Table 2). Table 5 also shows the average water consumption required (Wnc, m^3/month) for each type of household, which was calculated using Equation 3.

With regard to the water consumption required per month (Wnc) for each type of household, the maximum and minimum consumption of the ecological devices represent the following percentages with respect to the total consumption, respectively: (1) flush toilets (30.71% and 23.03%); (2) washing machine (8.79% and 7.99%); and (3) combined use (38.72% and 14.23%). Concerning the use of conventional flush toilet (6 L/flush) and laundry machine (120 L/load), the use of these ecological devices represents water savings up to 20% and 25% (maximum and minimum flush toilet volumes, respectively) and 97% and 43% (maximum and minimum washing machine volumes, respectively). In the case that rainwater harvesting at the household level was implemented, the water savings could contribute to alleviating the current problem of water scarcity that the population is facing.

Table 5. Maximum and minimum water consumption of ecological devices (Avc, m^3/month) depending on the type of household roof area (*Hra* = 45 m^2, 50 m^2, 100 m^2 and 200 m^2).

Water Consumption of Ecological Devices	45 m^2	50 m^2	100 m^2	200 m^2
Wnc (m^3/month)	14.07	15.59	17.11	19.39
Toilet, Avc$_{max}$	4.32	4.78	5.25	5.95
Toilet, Avc$_{min}$	3.24	3.59	3.94	4.46
Washing machine, Avc$_{max}$	1.23	1.37	1.50	1.70
Washing machine, Avc$_{min}$	1.12	1.24	1.36	1.55
Combined use, Avc$_{max}$	5.44	6.03	6.62	7.50
Combined use, Avc$_{min}$	2.00	2.21	2.43	2.76

The volume of rainwater harvestable monthly from the roof area (*VR*) was compared to the water consumption necessary for the flush toilet (toilet, Avc$_{max}$; and toilet, Avc$_{min}$), washing machine (washing machine, Avc$_{max}$; and washing machine, Avc$_{min}$), as well as for combined use (combined use, Avc$_{max}$; and combined use, Avc$_{min}$). Figures 5–7 depict the maximum and minimum consumption associated with the flush toilet, washing machine and the combined use of both ecological devices, respectively.

The balance between the rainwater harvesting volume collected from the roof area (*VR*) and the volumes of maximum consumption of flush toilets (toilet, Avc$_{max}$), indicate that during the months of December and January, there is no available water (*Wa*) in any type of household. For *Hra* of 200 m^2, there is *Wa* for other uses from February–November; for *Hra* of 100 m^2, there will be *Wa* only from May–October; while for *Hra* of 45 m^2 and 50 m^2, there will only be *Wa* for other uses than the flush toilet in the months of July, September and October (Figure 5).

It was also observed that during the months of November, January, February, March and April, the balance between the volume of rainwater harvested (*VR*) and the water used in the flush toilet (toilet, Avc$_{min}$) is negative, except for *Hra* of 200 m^2. For *Hra* of 100 m^2, there will be available water (*Wa*) from May to October; for *Hra* of 45 m^2 and 50 m^2, only during the months of July, September and October, there would be *Wa* after consumption from flush toilets (Figure 5).

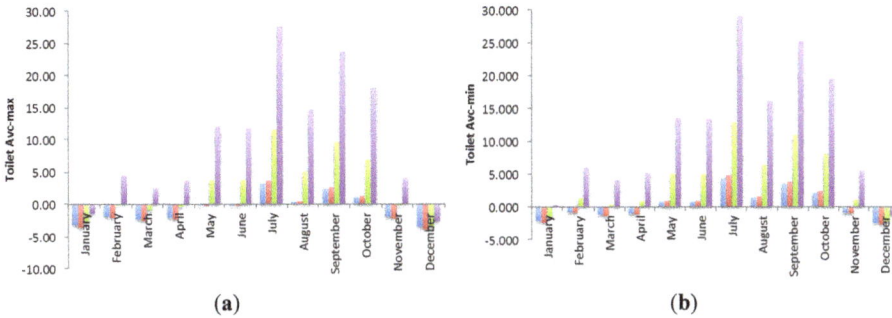

Figure 5. Monthly balance of rainwater per household type (available water (Wa) in m³/month) after flush toilet (a) Toilet, Avc_{max}; (b) Toilet, Avc_{min} blue: $Hra = 45$ m²; red: $Hra = 50$ m²; green: $Hra = 100$ m²; purple: $Hra = 200$ m²).

The maximum consumption for the washing machine (washing machine, Avc_{max}) for Hra of 200 m² could be covered the entire year. Households with Hra of 100 m², 50 m² and 45 m² could satisfy this need only from February to November. The months of December and January could be covered at 100% with the volume stored during the other months (Figure 6).

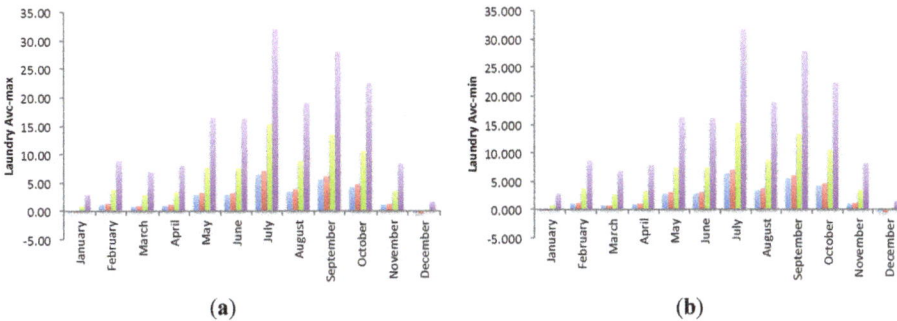

Figure 6. Monthly balance of rainwater per household type (Wa in m³/month) after washing machine consumption, (a) Laundry, Avc_{max}; (b) Laundry, Avc_{min} (blue: $Hra = 45$ m²; red: $Hra = 50$ m²; green: $Hra = 100$ m²; purple: $Hra = 200$ m²).

The household roof area (Hra) of 200 m² could satisfy the minimum consumption for the washing machine (washing machine, Avc_{min}) throughout the year, and it would provide a volume of available water (Wa) after the balance. The Hra of 100 m² would yield Wa all year, except in December. For Hra of 50 m² and 45 m², all of their water consumption could be covered from February to November; the Wa stored during these months would be sufficient to cover the demand during the months of December and January (Figure 6).

In the case of the maximum combined consumption (combined use, Avc_{max}), for Hra of 200 m², there would be an available water volume (Wa) from February to November; from May to October, volumes greater than 10 m³ could be stored, which would be enough to cover the volume necessary for the rest of the months. For Hra of 100 m², there would be Wa from May to October; with the volume stored during those months, the requirements of the five other months of the year could be met. For Hra of 50 m² and 45 m², only during the rainiest months (July and September) could volumes of 1.22 m³ and 2.09 m³ be stored, which could be used partially for at least another month (Figure 7).

In the case of minimum combined consumption (combined use, Avc_{min}) for Hra of 200 m², volumes (Wa) greater than 1 m³/month could be stored from January to November; while in the months of February, March, April and November, available volumes greater than 5 m^3/month could be obtained. In May and June, Wa is higher than 15 m^3/month, >30 m^3/month in July, >17 m^3/month in August, >25 m^3/month in September and >20 m^3/month in October. For *Hra* of 100 m^2, water could be stored from February to November (between 1 m^3/month and 14 m^3/month). For *Hra* of 45 m^2 and 50 m^2, there is *Wa* from May to October (between 2 m^3/month and 6 m^3/month, respectively), sufficient to cover other months of the year (Figure 7).

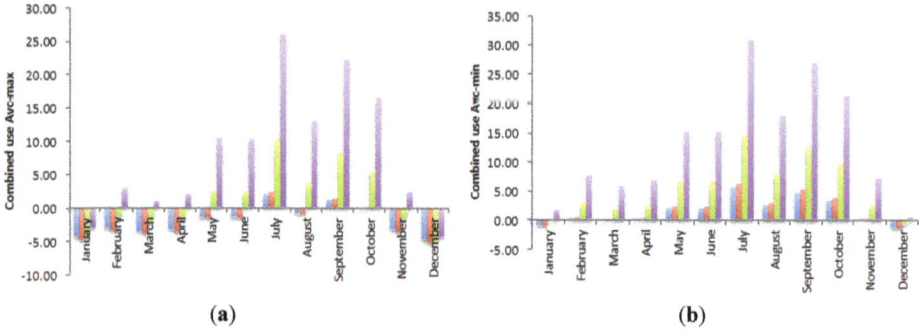

Figure 7. Monthly balance of rainwater per household type (*Wa* in m^3/month) after combined use (a) Combined use, Avc$_{max}$; (b) Combined use, Avc$_{min}$ (blue: *Hra* = 45 m^2; red: *Hra* = 50 m^2; green: *Hra* = 100 m^2; purple: *Hra* = 200 m^2).

4.3. Discussion

The harvesting of rainwater through its interception on rooftops represents an important option to take advantage of rainfall in places facing problems of water stress. In the study area of Pachuca-Mineral de la Reforma, with an average annual rainfall of 585.60 mm, the results obtained indicated the following:

If *VR* is utilized only for the maximum consumption of flush toilets (toilet, Avc$_{max}$), only in households with *Hra* of 100 m^2 and 200 m^2 would the needs all year round be met. If the consumption of the flush toilet is considered with the minimum volumes quantified (toilet, Avc$_{min}$), the four types of household analyzed cover 100% of their necessary consumption. In addition, there would be available water volumes (*Wa*) from 1 m^3/month up to 135 m^3/month to be used in other non-drinking uses throughout the year.

If the rainwater collected is solely used for laundry (washing machine, Avc$_{max}$; washing machine, Avc$_{min}$), 100% of the consumption of any type of household would be covered. In addition, there would be water volumes (*Wa*) from 28.9 m^3/month to 171 m^3/month, depending on the *Hra*, available for other non-potable purposes during the year.

On the other hand, if a maximum combined consumption (combined use, Avc$_{max}$) in flush toilets and washing machines is chosen in homes with *Hra* of 45 m^2 and 50 m^2, Wa to be stored during the period of drought only would cover 12.7% and 12.9%, respectively, of the necessary consumption in those months. In households with *Hra* of 100 m^2, the *Wa* obtained during the rainy season is enough to cover the needs for the rest of the year, and there would be a *Wa* of 15 m^3/month useful for other non-drinking purposes. Regarding the households with *Hra* of 200 m^2, besides covering the maximum combined consumption (combined use, Avc$_{max}$) there would be spare available *Wa* of 98.55 m^3/month, possibly enough to cover other non-drinking purposes throughout the year.

If a combined minimum consumption is selected (combined use, Avc$_{min}$), any type of *Hra* analyzed would cover its consumption throughout the year. In addition, they would have *Wa* from 18 m^3/month (*Hra* of 45 m^2) to 156 m^3/month (*Hra* of 200 m^2). In the latter type of household, with such an available volume, it is advisable to install a purification system, which would make this type of home almost

Water **2015**, *7*, 4622–4637

completely independent of the public potable water supply. However, if rainwater harvesting is implemented in order to satisfy at least the volume consumption required for a combined use in flush toilets and washing machines, this alternative could be sufficient to reduce the problem of water shortage from which the population of the study area suffers.

5. Conclusions

This study confirms that rainwater harvesting on household roof areas is a viable option, even in arid areas, such as that studied in this paper, where the average annual rainfall is 585.6 mm.

By estimating the daily and monthly rainfall, it was possible to determine that harvestable volumes are sufficient to meet flush toilet consumption, washing machines, as well as their combined use throughout the year, depending on the household roof area (*Hra*). However, if daily or hourly rainfall data were used, these results could provide more accurate interpretations that would help to quantify the storage tank with reliability.

However, households with roof areas of (*Hra*) 45 m^2 and 50 m^2 still depend on the water supply system of the public network during part of the year. On the other hand, in households with roof areas (*Hra*) of 100 m^2 and 200 m^2, besides covering the consumption of flush toilets and washing machines throughout the year, harvested rainwater still might be stored to meet other needs. Even the drinking use of this water is feasible, but proper treatment systems must be installed.

Therefore, it can be concluded that it is possible to establish indicators that help users to estimate the minimum capacity necessary for installing a storage system in their homes. In addition, it is recommended that the users take into account that, for the installation of the storage system, they must have filters that separate the organic matter and the dust that may accumulate on the roof area. In the case of the *Hra* of 100 m^2 and 200 m^2, it is also recommended to consider commercial systems for drinking water purification at the household level.

Acknowledgments: The first author thanks Carlos Alexander Lucho Constantino for the descriptive statistical analysis. The authors thank the four anonymous reviewers for their timely review and helpful comments.

Author Contributions: Liliana Lizárraga-Mendiola proposed the conceptual framework of the paper, carried out the data analysis and prepared the first draft of the paper. Gabriela Vázquez-Rodríguez checked the first draft of the paper and results and enhanced the discussion and writings of the paper. Alberto Blanco-Piñón reviewed bibliography related with rainwater harvesting. Yamile Rangel-Martínez analysed the national household classification and water consumption. María González-Sandoval reviewed the methods proposed and the final version of the paper.

Conflicts of Interest: The authors declare no conflict of interest.

References

1. Willuweit, L.; O'Sullivan, J.J. A decision support tool for sustainable planning of urban water systems: Comes across the dynamic urban water simulation model. *Water Res.* **2013**, *47*, 7206–7220. [CrossRef] [PubMed]
2. Belmeziti, A.; Coutard, O.; de Gouvello, B. A new methodology for evaluating potential for potable water savings (PPWS) by using rainwater harvesting at the urban level: The case of the municipality of Colombes (Paris Region). *Water* **2013**, *5*, 312–326. [CrossRef]
3. African Development Bank. *Rainwater Harvesting Handbook: Assessment of Best Practices and Experience in Water Harvesting*; African Development Bank: Abidjan, Côte d'Ivoire, 2008.
4. Sapkota, M.; Arora, M.; Malano, H.; Moglia, M.; Sharma, A.; George, B.; Pamminger, F. An overview of hybrid water supply systems in the context of urban water management: Challenges and opportunities. *Water* **2015**, *7*, 153–174. [CrossRef]
5. Liaw, C.H.; Chiang, Y.C. Dimensionless analysis for designing domestic rainwater harvesting systems at the regional level in northern Taiwan. *Water* **2014**, *6*, 3913–3933. [CrossRef]
6. Villarreal, E.L.; Dixon, A. Analysis of a rainwater collection system for domestic water supply in Ringdansen, Norrköping, Sweden. *Build. Environ.* **2005**, *40*, 1174–1184. [CrossRef]

7.   Khastagir, A.; Jayasuriya, N. Optimal sizing of rain-water tanks for domestic water conservation. *J. Hydrol.* **2010**, *381*, 181–188. [CrossRef]
8.   Imteaz, M.A.; Adeboye, O.B.; Rayburg, S.; Shanableh, A. Rainwater harvesting potential for southwest Nigeria using daily water balance model. *Resour. Conserv. Recycl.* **2012**, *62*, 51–55. [CrossRef]
9.   Imteaz, M.A.; Ahsan, A.; Shanableh, A. Reliability analysis of rainwater tanks using daily water balance model: Variations within a large city. *Resour. Conserv. Recycl.* **2013**, *77*, 37–43. [CrossRef]
10. Rahman, A.; Keane, J.; Imteaz, M.A. Rainwater harvesting in greater Sydney: Water savings, reliability and economic benefits. *Resour. Conserv. Recycl.* **2012**, *61*, 16–21. [CrossRef]
11. Su, M.D.; Lin, C.H.; Chang, L.F.; Kang, J.L.; Lin, M.C. A probabilistic approach to rainwater harvesting systems design and evaluation. *Resour. Conserv. Recycl.* **2009**, *53*, 393–399. [CrossRef]
12. Jones, M.P.; Hunt, W.F. Performance of rainwater harvesting systems in the Southeastern United States. *Resour. Conser. Recycl.* **2010**, *54*, 623–629. [CrossRef]
13. United Nations Development Programme (UNDP). *Pilot Deployment of Harvesting of Rain-Water, as a Measure of Adaptation to Climate Change*; University of Guanajuato: Guanajuato, Mexico, 2013.
14. Lluvia Para Todos: Por un Mexico Sustentable en Agua. Available online: http://islaurbana.mx/contenido/ biblioteca/presentaciones/IslaUrbana-CaptaciondeLluvia.png (accessed on 10 March 2015). (In Spanish)
15. Water Management in the Metropolitan Area of Mexico City: The Hard Way to Learn. Available online: http://hydrologie.org/ACT/Marseille/works-pdf/wchp5-3.png (accessed on 3 April 2015).
16. Mazari-Hiriart, M. Water Latin America. Inter-American Association of Sanitary and Environmental Engineering (AIDIS). Available online: http://www.preventionweb.net/english/professional/contacts/v. php?id=3650 (accessed on 23 June 2015).
17. Salazar-Adams, A.; Pineda-Pablos, N. Factors that affect the water demand for domestic use in Mexico. *Reg. Soc.* **2010**, *49*, 3–16.
18. Estadísticas del Agua en México. Available online: http://www.conagua.gob.mx/CONAGUA07/ Publicaciones/Publicaciones/EAM2014.png (accessed on 2 February 2015). (In Spanish).
19. Neri-Ramírez, E.; Rubiños-Panta, J.E.; Palacios-Velez, O.L.; Oropeza-Mota, J.L.; Flores-Magdaleno, H.; Ocampo-Fletes, I. Evaluation of the sustainability of the aquifer Cuautitlan-Pachuca through the use of the MESMIS methodology. *Chapingo Mag.* **2013**, *19*, 273–285.
20. Galindo-Castillo, E.; Otazo-Sanchez, E.M.; Gordillo-Martínez, A.J.; Arellano-Islas, S.; González-Ramírez, C.A.; Reyes-Gutierrez, L.R. Impact, technology and environmental toxicology. In *Depletion of the Aquifer Cuautitlan-Pachuca: Water Balance 1990–2010*; Universidad Autónoma del Estado de Hidalgo: Pachuca, Mexico, 2011.
21. Amaya-Ventura, M.L. Institutional aspects of water management in Pachuca, Hidalgo. *Mex. J. Soc.* **2011**, *73*, 509–537.
22. Huizar-Álvarez, R. Hydrogeological map of the Río de las Avenidas basin in Pachuca, Mexico. *Invest. Geogr. Geol. Inst. Bull.* **1993**, *27*, 95–131.
23. National Water Comission-Meteorological National Service. Available online: http://smn.cna.gob.mx (accessed on 8 December 2014).
24. Consejo Estatal de Población de Hidalgo. Available online: http://www.conapo.gob.mx/es/CONAPO/ Hidalgo (accessed on 31 January 2015). (In Spanish)
25. Comisión De Agua Y Alcantarillado De Sistemas Intermunicipales. Available online: http://caasim.hidalgo. gob.mx (accessed on 12 November 2014). (In Spanish)
26. Aladenola, O.O.; Adeboye, O.B. Assessing the potential for rainwater harvesting. *Water Resour. Manag.* **2010**, *24*, 2129–2137. [CrossRef]
27. Ahammed, F.; Alankarage-Hewa, G.; Argue, J.R. Variability of annual daily maximum rainfall of Dhaka, Bangladesh. *Atmos. Res.* **2014**, *137*, 176–182. [CrossRef]
28. United Nations Environment Programme. Potential for Rainwater Harvesting in Africa: A GIS Overview. Available online: http://www.unep.org/pdf/RWH_in_Africa-final.png (accessed on 8 January 2015).
29. Secretaria de Desarrollo Agrario, Territorial Y Urbano. Available online: http://www.conavi.gob.mx/ images/documentos/plan_nacional_desarrollo_2013/2014/programa_nacional_de_vivienda_2014-2018. png (accessed on 2 December 2014). (In Spanish)

30. National Institute of Ecology and Climatic Change. Sustainable Household. Habits and Bathroom Technologies. Available online: http://www.inecc.gob.mx (accessed on 8 January 2015).
31. Anaya-Garduño, M. *Rainwater Harvesting. Solution Fall from Heaven*; National Institute of Ecology and Climatic Change: Mexico City, Mexico, 2011.

water

MDPI

Article

Optimal Spatial Design of Capacity and Quantity of Rainwater Harvesting Systems for Urban Flood Mitigation

Chien-Lin Huang [1], Nien-Sheng Hsu [1,*], Chih-Chiang Wei [2] and Wei-Jiun Luo [1]

[1]   Department of Civil Engineering, National Taiwan University, No. 1, Sec. 4, Roosevelt Road,
      Taipei 10617, Taiwan; d98521008@ntu.edu.tw (C.-L.H.); madmichae@hotmail.com (W.-J.L.)
[2]   Department of Marine Environmental Informatics, National Taiwan Ocean University, No. 2, Beining Road,
      Jhongjheng District, Keelung City 20224, Taiwan; d89521007@ntu.edu.tw
*   Author to whom correspondence should be addressed; nsshue@ntu.edu.tw; Tel.: +886-2-3366-2640;
      Fax: +886-2-3366-5866.

Academic Editor: Ataur Rahman
Received: 8 July 2015; Accepted: 15 September 2015; Published: 23 September 2015

Abstract: This study adopts rainwater harvesting systems (RWHS) into a stormwater runoff management model (SWMM) for the spatial design of capacities and quantities of rain barrel for urban flood mitigation. A simulation-optimization model is proposed for effectively identifying the optimal design. First of all, we particularly classified the characteristic zonal subregions for spatial design by using fuzzy C-means clustering with the investigated data of urban roof, land use and drainage system. In the simulation method, a series of regular spatial arrangements specification are designed by using statistical quartiles analysis for rooftop area and rainfall frequency analysis; accordingly, the corresponding reduced flooding circumstances can be simulated by SWMM. Moreover, the most effective solution for the simulation method is identified from the calculated net benefit, which is equivalent to the subtraction of the facility cost from the decreased inundation loss. It serves as the initially identified solution for the optimization model. In the optimization method, backpropagation neural network (BPNN) are first applied for developing a water level simulation model of urban drainage systems to substitute for SWMM to conform to newly considered interdisciplinary multi-objective optimization model, and a tabu search-based algorithm is used with the embedded BPNN-based SWMM to optimize the planning solution. The developed method is applied to the Zhong-He District, Taiwan. Results demonstrate that the application of tabu search and the BPNN-based simulation model into the optimization model can effectively, accurately and fast search optimal design considering economic net benefit. Furthermore, the optimized spatial rain barrel design could reduce 72% of inundation losses according to the simulated flood events.

Keywords: rainwater harvesting system; stormwater runoff management model; backpropagation neural network; tabu search; spatial design of capacity and quantity; optimization; urban flood mitigation

---

1. Introduction

In recent years, on account of global climate change and the increasing occurrence of extreme hydrological events, coupled with the fact that Taiwan is densely populated and overdeveloped in catchment areas, the amount of flooding caused by heavy rain often exceeds the scale of the originally designed standard. Additionally, the drainage system in Taiwan is insufficient, which causes the water level to rise extremely quickly during typhoons and heavy rainfall. The pumping station of the urban drainage system cannot handle such large amount of flood in recent years, this leads to flooding and the subsequent loss of life and property. In response to this challenging situation, new modes, measures and solutions should be developed to achieve the goal and evaluate the feasibility for flood mitigation.

Low-impact development (LID) provides techniques for innovative urban environmental planning, management, and environmental protection. The frequently used techniques include rain barrels, green roofs, permeable paving, roadside ecological spaces, rainwater harvesting systems (RWHS), and others. The LID facilities have relatively lower costs in reducing peak and total runoff compared to traditional flood control measures for building underground pipeline culverts. Moreover, LID facilities can provide additional benefits, such as water conservation, urban beautification, and improvement of the ecological environment. Among these facilities, RWHS can be implemented on in-place water harvesting, which differs from the traditional drainage concept of the end-trace centralization process. RWHS are containers that collect roof runoff during storm events and can either release or re-use the rainwater during dry periods. RWHS collect runoff from rooftops and convey it to a cistern tank. Furthermore, RWHS are easy to obtain, cause less pollution and costs at a lower risk, and involve no water right disputes. In short, these systems can serve as flood detention means and alternative water sources that are worthy of broad use.

Previous studies regarding RWHS can be divided into two categories. The first one is the capacity design of RWHS under the consideration of domestic water supply those primarily employ a simulation method for planning. The related studies are described as follows: Liaw and Tsai (2004) [1] developed a simulation model including production to estimate the most cost effective combination of the roof area and the storage capacity that best supplies a specific volume of water. Liaw and Chiang (2014) [2] developed a regional-level and dimensionless analysis for designing a domestic RWHS. Moreover, regarding design using economic and dimensionless analysis-based optimization approach, Chiu *et al.* (2009) [3] optimized the most cost-effective rainwater tank volumes for different dwelling types using marginal analysis. Campisano and Modica (2012) [4] developed a dimensionless methodology for the optimal design of domestic RWHS. From these studies, we can find out that previous studies have scarcely designed the capacity of RWHS considering flood reduction benefits using an interdisciplinary integrated systematic analysis approach. In addition, the capacity design of RWHS is primarily limited to small communities and lacks full consideration of all metropolitan catchment with variations in spatial capacity and quantity design of RWHS.

The second category regarding RWHS is simulation and evaluation of the effectiveness and reliability of domestic RWHS with a variety of patterns on the water supply objects. The research subjects include: (1) evaluating the potential for potable water savings by using rainwater in residential sectors [5,6]; (2) estimating nonpotable household potential, sustainability and performance of storage type of RWHS [7–9] and investigating the potential benefits from sharing RWHS with nearby neighbors with a storage-reliability-yield analysis (Seo *et al.*, 2015 [10]) using rainfall data; and (3) establishing the probabilistic relationships between storage capacities and deficit rates of RWHS [11] and that of between the efficiency of rural domestic rainwater management and tank size, tank operation and maintenance, respectively [12]. However, these studies seldom consider the surface and sewer physical flowing phenomenon after rainwater partially intercepted by RWHS and partially flowing to ground and urban drainage system. To address these problems, there are numerous studies evaluating and assessing the performance and reliability of RWHS using numerical or hydrological model. The related studies are, for example, Jones and Hunt (2010) [13] evaluated the performance of RWHS by a monitoring study with a computer model (Rainwater Harvester 3.0), Basinger *et al.* (2010) [14] assessed the reliability of RWHS using a novel model based on a nonparametric rainfall generation procedure utilizing a bootstrapped Markov chain, and Palla *et al.* (2011) [15] proposed nondimensional parameters with a suitable behavioral model according to a daily mass balance equation to investigate optimum performance of RWHS. However, these studies almost only estimated the efficiency of RWHS for nonpotable household water saving that did not assess the feasibility for flood mitigation. In addition, the performance of RWHS for stormwater retention has been studied, such as [16–18].

However, these studies seldom simulated, evaluated and account for the inundated loss of each actual flood event in terms of the space design patterns of RWHS.

The purpose of this study is to develop a set of novel simulation-optimization models to identify the most effective spatial design for a quantity and capacity arrangement of RWHS in urban drainage areas considering fast and effective optimization of flooding loss reduction and facility cost minimization. The effective characteristic zonal subregions for spatial design are particularly classified by using fuzzy C-means (FCM) clustering with the investigated data of urban roof, land use and rainfall characteristic among drainage area, and a series of representative regular spatial arrangements specification are designed by using statistical quartiles analysis for rooftop area and rainfall frequency analysis. A backpropagation neural network-based [19,20] water level simulation model is embedded in the optimization model, and used to substitute for the hydrologic/hydraulic-based storm water management model [21,22] to conform to newly considered interdisciplinary multiobjective optimization model, and combine it with tabu search (Glover, 1986; Glover and Laguna, 1997 [23,24]) to achieve the optimization process.

2. Development of Methodology

2.1. Procedures

The methodology of this study is divided into two parts: a simulation method and a hybrid simulation-optimization method. Information obtained from the simulation method is entered into the optimization model to produce the optimal solution. The flowchart of the methodology can be shown in Figure 1, and the steps are described as follows.

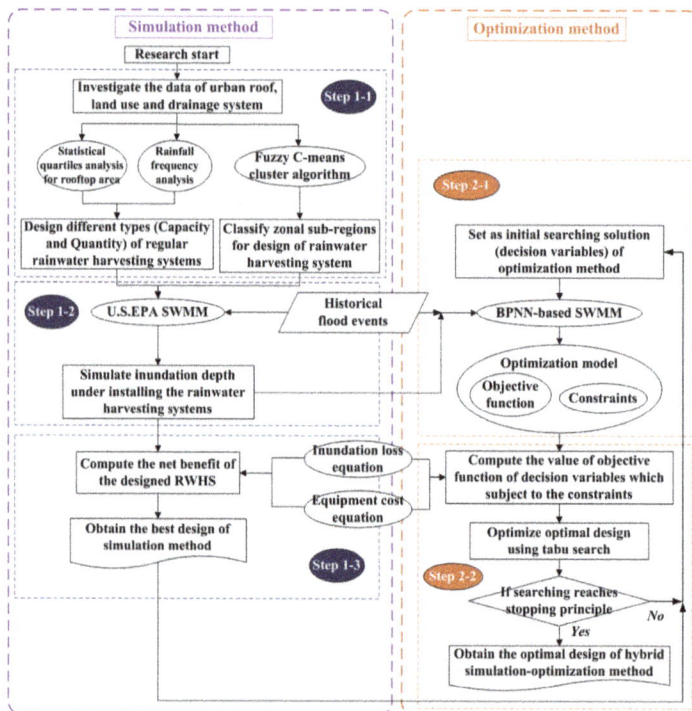

Figure 1. Flowchart of the methodology.

Step 1-1: Investigate the data of urban roof, land use and drainage system. Then design the regular spatial quantity and capacity arrangement of different types of RWHS in SWMM by using statistical quartiles analysis for rooftop area and rainfall frequency analysis, and classify zonal subregions for design of RWHS by using FCM cluster algorithm.

Step 1-2: Input the actual storm events to the constructed SWMM model to simulate the flooding and water level of the control points for different spatial RWHS designs and rain types.

Step 1-3: Convert the flooding amount into inundation loss and subtract the equipment cost to simulate the net benefit with various types of designs, and then obtain the best RWHS design of simulation method.

Step 2-1: Devise a suitable solution obtained from the simulation method as initial searching solution of the optimization method. Establish a water level simulation model for the urban drainage system that can substitute for U.S. EPA SWMM using time sequence data obtained from the simulation method with the BPNN. Then, the BPNN-based water level simulation model is embed into the defined optimization model which is composed of an objective function and constraints.

Step 2-2: Employ a tabu search algorithm to obtain the optimal solution of the optimization method, and then obtain the excellent spatial design of RWHS considering the urban flood reduction benefits.

2.2. Development of Simulation Model for Spatial Arrangement of Quantity and Capacity

This study outlines various specifications for the rain barrel spatial distribution and quantity design approach and applies SWMM to simulate the burst pipes and flooding situation for each case in numerous rainstorm events. The regular spatial quantity and capacity arrangement of different types of RWHS are designed by using statistical quartiles analysis for rooftop area and rainfall frequency analysis, and the zonal subregions for design of RWHS are classified by using FCM cluster algorithm.Moreover, it estimates the economic net benefit. The design patterns for various cases involve (1) rain barrels distributed throughout the entire region; (2) concentration on the downstream of the flooding region; and (3) concentration on the upstream of the flooding region. The detailed developed methodology is described in the following.

2.2.1. Classified Methodology of Zonal Subregions for Design of Rainwater Harvesting System

In an urban drainage area, the spatial distribution of building rooftop area and terrain is highly divergent and complex. The available rooftop material for installing RWHS is the surface which directly receives the rainfall and provides water to the system. It can be a paved area like a terrace or courtyard of a building, or an unpaved area like a lawn or open ground. A roof made of reinforced cement concrete (RCC), galvanized iron or corrugated sheets can also be used for water harvesting. Besides, the efficiencies of actual water storage in an identical rainfall cluster can approximately reflect a specific range with fewer variations because of the similarity of rainfall intensity and duration [25], so the average precipitation is also devised as designed basis. In order to reduce unnecessary searching solution space and be convenient for effective urban planning, this study applies FCM cluster algorithm to classify the study area to characteristic zonal subregions. The building region with similar geophysical characteristic of rooftop area, terrain height and rainfall will be clustered into same subregions. The FCM algorithm [26] is one of the most widely used fuzzy clustering algorithms. The FCM algorithm attempts to partition a finite collection of n elements $X = \{x_1, x_2, ..., x_n\}$ (x_i is set as a vector of rooftop area, terrain height and average precipitation in this study) into a collection of fuzzy clusters with respect to some given criterion. Given a finite set of data, the algorithm returns a list of c cluster centers $C = \{c_1, c_2, ..., c_c\}$ and a partition weighting matrix $W = w_{ij} \in [0, 1]$, $i = 1, 2, ..., n, j = 1, 2, ..., c$, where each element $w_{i,j}$ tells the degree to which element

x_i belongs to cluster c_j. The FCM aims to minimize an objective function (J) which is expressed as follows:

$$Min \ J = \sum_{i=1}^{n} \sum_{j=1}^{c} w_{ij}^m \|x_i - c_j\|^2 \qquad m \in R \quad \cap \quad m \geq 1 \tag{1}$$

where partition weighting matrix (w_{ij}) and cluster centers (c_j) can be calculated using the following Equations:

$$w_{ij} = \frac{1}{\sum_{k=1}^{c} \left(\frac{\|x_i - c_j\|}{\|x_i - c_k\|} \right)^{\frac{2}{m-1}}} \tag{2}$$

$$c_j = \frac{\sum_{i=1}^{n} w_{ij}^m x_i}{\sum_{i=1}^{n} w_{ij}^m} \left| \begin{array}{l} 0 \leq w_{ij} \leq 1 \\ \sum_{j=1}^{c} w_{ij} = 1 \end{array} \right. \tag{3}$$

2.2.2. Design Methodology of Capacity and Quantity of Regular Rainwater Harvesting Systems

In an urban drainage area, the design principle of RWHS for flood mitigation is to store storm rainwater as much as possible to maximize economic urban flood reduction benefits while the designed specification must be subject to the limitation of available building rooftop area. The designed parameters for RWHS include capacity (volume) and quantity (arranged density). This study invents an approach to generate a series of representative regular spatial capacity and quantity arrangements of RWHS. The volume of rain barrel (S_r) is specialized as available design area (A_l) multiplying to rainfall intensity of target desired stored precipitation of the specific return periods ($P_T^{\hat{R}P}$) (Equation (4)), in order to mitigate the heavy rains induced flood. The variable $P_T^{\hat{R}P}$ can be evaluated by rainfall frequency analysis using the probability distribution of normal, log–normal, extreme-value type I, Pearson type III or log-Pearson type III (adopted by this study; Lee and Ho, 2008 [27]). In addition, the arranged density is set as how many areas arrange one rain barrel in SWMM. Hence, it is a key factor to determine the representational arranged area which can be subject to the lowest and highest limitation in practical urban buildings. This study applies statistical quartiles analysis with investigated spatial rooftop area to determine the representational arranged area (Equation (5)).

$$S_r = \left[A_l \cdot P_T^{\hat{R}P} \left| \begin{array}{l} T \in [1, 2, ..., D] \\ l \in [1, 2, ..., L] \end{array} \right. \right] \tag{4}$$

$$A_l = [Min(a_r^{min}), WA(a_r^{min}), WA(a_r^{med}), Min(a_r^{q\%}), WA(a_r^{q\%})] \tag{5}$$

where a_r^{min}, a_r^{med}, and $a_r^{q\%}$ are minimum, medium and q percentage of quartiles rooftop area on subregion r, respectively; and WA means weighted average.

2.2.3. Assessment Index of Designed Goodness

This study identifies the annual net benefit after establishing the RWHS as an indicator to evaluate the flooding reduction effect of different design approaches. The annual net benefit is the average annual flooding loss reduction minus the annual cost. This reduction is derived from the flooding loss without employing the RWHS design approach minus the flooding loss with employing it. The annual cost is the average annual setup cost of the RWHS.

2.2.4. Computation of Inundation Loss

In practice, the flooding loss is directly proportional to inundated depth which is directly proportional to total volume of burst pipes. SWMM can calculate the burst pipe amount (*i.e.*, volume) in the manhole at each point in time through the simulation. Moreover, the spatial-temporal flooding scope and depth can also be calculated by the temporal-spatial burst pipe volume with the volume-depth-width relationship in the inundation region. The calculation of flooding loss can be divided into residential and commercial districts. Accordingly, we can calculate the flooding loss using the characteristic curve equation constructed by investigated data that the evaluated factor is total volume of burst pipes (use 2 term polynomial function as example):

$$[L_p^{\text{non}}, L_p^{\text{RWHS}}] = b_0 + b_1[F_p^{\text{total-non}}, F_p^{\text{total-RWHS}}] + b_2[F_p^{\text{total-non}}, F_p^{\text{total-RWHS}}]^2 \tag{6}$$

$$[F_p^{\text{total-non}}, F_p^{\text{total-RWHS}}] = \left[\sum_{t=1}^{T} F_p^{\text{non}}(t), \sum_{t=1}^{T} F_p^{\text{RWHS}}(t)\right] \tag{7}$$

where $F_p^{\text{total-non}}$ and $F_p^{\text{total-RWHS}}$ is the total volume of burst pipes in the flooded areas at control point p with no RWHS design and with RWHS design, respectively; and $F_p^{\text{non}}(t)$ and $F_p^{\text{RWHS}}(t)$ is the volume of burst pipes at moment t and control point p with no RWHS design and with RWHS design, respectively.

2.3. Introduction of SWMM

The United States Environmental Protection Agency (US EPA) SWMM model is a dynamic rainfall–runoff simulation model used for single-event to long-term (continuous) simulation of the surface/subsurface hydrology quantity and quality from primarily urban/suburban areas [21,22]. The hydrology component of SWMM operates on a collection of subcatchment areas with and without depression storage to predict runoff from precipitation, evaporation and infiltration losses from each of the subcatchment. In addition, the LID areas on the subcatchment can be modeled to reduce the impervious and pervious runoff. SWMM tracks the flow rate, flow depth, and water quality in each pipe and channel during a simulation period composed of multiple fixed or variable time steps. In the simulations, the runoff component of SWMM (RUNOFF) operates on a collection of subcatchment areas that receive precipitation and generate runoff. The routing portion of SWMM transports this runoff through a system of pipes, channels, storage/treatment devices, pumps, and regulators.

2.3.1. Model Parameters and Routing

The adopted model parameters for simulation in subcatchments are surface roughness, depression storage, slope, flow path length; for Infiltration, is Horton-based max/min rates and decay constant; and for Conduits, is Manning's roughness. A study area can be divided into any number of individual subcatchments, each of which drains to a single point. The subcatchment width parameter is normally estimated by first estimating a representative length of overland flow, then dividing the subcatchment area by this length. Ideally this should be the length of sheet flow (<100 m), which is typically significantly slower than channelized flow.

The routing options of SWMM include steady flow routing, kinematic wave routing and dynamic wave routing. Dynamic wave routing solves the complete one-dimensional Saint-Venant flow equations and therefore produces the most theoretically accurate results. These equations consist of the continuity and momentum equations for conduits and a volume continuity equation at nodes. With this form of routing it is possible to represent pressurized flow when a closed conduit becomes full, such that flows can exceed the full normal flow value. The excess flow is either lost from the system or can pond atop the node and re-enter the drainage system. Dynamic wave routing can account for channel storage, backwater, entrance/exit losses, flow reversal, and pressurized flow, because it couples together the solution for both water levels at nodes and flow in conduits. Due to the capability and demand of this study, dynamic wave routing is applied for routing.

2.3.2. The Rainwater Harvesting Function within Low-impact Development Components

The LID function is integrated within the subcatchment component of SWMM and allows further refinement of the overflows, infiltration flow and evaporation in rain barrel, swales, permeable paving, green roof, rain garden, bioretention and infiltration trench. LID takes many forms but can generally be thought of as an effort to minimize or prevent concentrated flows of storm water leaving a site [28]. The RWHS is one of the LID techniques in SWMM, and the RWHS is assumed to consist of a given number of fixed-sized cisterns per 1000 ft^2 (or 90 m^2) of rooftop area captured.

2.4. Development of Optimization Model

In the developed optimization model of this study for the optimal spatial arrangement and capacity design of RWHS, the objective function is devised as the optimized annual net benefit. The constraints include the upper and lower limits of the rain barrel capacity and quantity, estimate equation of the rain barrel annual cost, equation converting the burst pipe volume into flooding loss, and BPNN drainage system simulation equation, among others.

This study generates data from the fully constructed SWMM combined with the simulation method. Furthermore, we utilize the BPNN model to substitute for US EPA SWMM to conform to newly considered interdisciplinary multi-objective optimization model and embed it into the optimization model. The purpose is to quickly and effectively produce optimal solutions with no limit of multi-embedded interdisciplinary simulation model. The formulaic descriptions of the optimization model are described below.

2.4.1. Objective Function

The objective function of RWHS optimization model is annual net benefit that is equal to the flooding loss deduction minus the facility cost of RWHS. The greater the value, the better it is; nevertheless, because the objective function aims to obtain the minimum value, we use the objective function to maximize the annual net benefit and obtain the minimum negative net benefit:

$$Min \ Z = -\left\{ \left[L_p^{non} - \sum_{p=1}^{P} L_p^{RWHS}(N_r, S_r) \Big|_{r \in [1,2,...,R]} \right] - \sum_{r=1}^{R} [T_c(S_r) \times N_r] \right\} \tag{8}$$

where T_c is the cost of the RWHS, N_r is the number of rain barrels in subregion r, L_{non} represents the flooding loss with no rain barrels established, and L_p is the flooding loss at control point p. In addition, S_r denotes the capacity of the rain barrel in subregion r, R is the quantity of the total subregion, and P represents the quantity at the flooding control point, wherein decision-making variables are the quantity of rain barrels in each N_r and S_r.

2.4.2. Constraints

(1) Quantity Constraints of Rain Barrels

To ensure that the quantity of rain barrels does not exceed the possible maximum quantity of each design subcatchment in each household that the quantity arrangement must meanwhile be subject to physical constraints, it is necessary to define the upper and lower limits of the quantity of rain barrels. The constraints can be expressed as:

$$0 \leq N_r \leq N_r^{max} \tag{9}$$

where N_r^{max} is the maximum quantity of rain barrels in subregion r.

(2) Capacity Constraints of Rain Barrels

To ensure that the capacity of rain barrels aligns with the physical size limitations without exceeding the upper and lower limits of the rain barrel design, we set the constraints as:

$$0 \leq S_r \leq S_r^{max} \tag{10}$$

where S_r^{max} represents the maximum capacity of rain barrels in subregion r that can be estimated by investigated available rooftop design area multiplying to rainfall intensity of target desired stored precipitation of the set maximum possible return periods (Section 2.2.2).

(3) Annual cost function of rainwater harvesting systems

In this study, the RWHS cost calculation employs the cost equation proposed by Liaw and Tsai (2004) [1], which was obtained through market research and analysis. This annual cost equation can be described as:

$$T_c(C_a, A_r) = a + bC_a^2 + cA_r \tag{11}$$

where $T_c(C_a, A_r)$ is the cost of RWHS, and function of the capacity C_a (m^3) and the roof area A_r (m^2).

(4) Equation for Transferring Flooding to Inundation Loss

In this study, flooding loss is converted from the total volume of burst pipes with a relationship characteristic curve equation (Equations (6) and (7)).

(5) Routing Equation of Drainage System

The water level calculation of the drainage system uses BPNN to construct an alternative simulation model. The routing equation at each control point can be described as:

$$H_p(t) = f\left(net_j^i(t)\right) \tag{12}$$

where $H_p(t)$ is the water level of the drainage system at control point p at time t; and f is the activation function. The accumulated weight value of the $n - 1$-th layer output value $net_j^i(t)$ is calculated by:

$$net_j^i(t) = \sum_i w_{ij}^n y_i^{n-1}(t) - b_j^n \tag{13}$$

where w_{ij}^n represents the connection weights of the n-th layer j-th neuron and $n - 1$-th layer i-th neuron. In addition, b_j^n denotes the bias weights of the n-th layer j-th neuron. $y_i^{n-1}(t)$ is the input variable of the model, which includes the precipitation, water level, and capacity and quantity of rain barrels in each subregion.

2.4.3. Solution of Optimization Model

We employ tabu search to select the optimal solution for the spatial arrangement and capacity design of RWHS under consideration of the benefits of urban flood reduction. Tabu search is widely applied to management and planning issues; it can efficiently identify nonlinear or optimal solutions. Furthermore, it can be easily combined with the optimization model, and it quickly and automatically selects the best solution for the decision-making variables. Besides, the decision variables of this study include zonal capacity and quantity of RWHS, and the quantity must be a natural number. A most important advantage of tabu search is that the searching moving distance can be set as integer, so this study selects tabu search as an optimization algorithm.

(1) Tabu Search

Proposed by Glover (1986) [23] and Glover and Laguna (1997) [24], tabu search guides the search direction and region using different types of memory. During the search, a search direction or region can be favored or prohibited according to the memory and rules. Additionally, the search can exit a local optimum region and avoid repeated searches through the definition of a tabu list, which includes the type and length of the search variables and associated objective function values. In each iteration, it only searches to find the best candidate solution. Hence, this search mechanism can significantly improve the search efficiency and accuracy and obtain the best global solution.

(2) Optimizing the Spatial Design of Quantity and Capacity by Tabu Search

We use tabu search to select the optimal rain barrel spatial arrangement and capacity design in each flood event; its flowchart is shown in Figure 2. The selection method sets the decision-making variables in the optimization model—*i.e.*, the quantity and capacity of the rain barrels in each region—as the tabu search solution. The steps are described below.

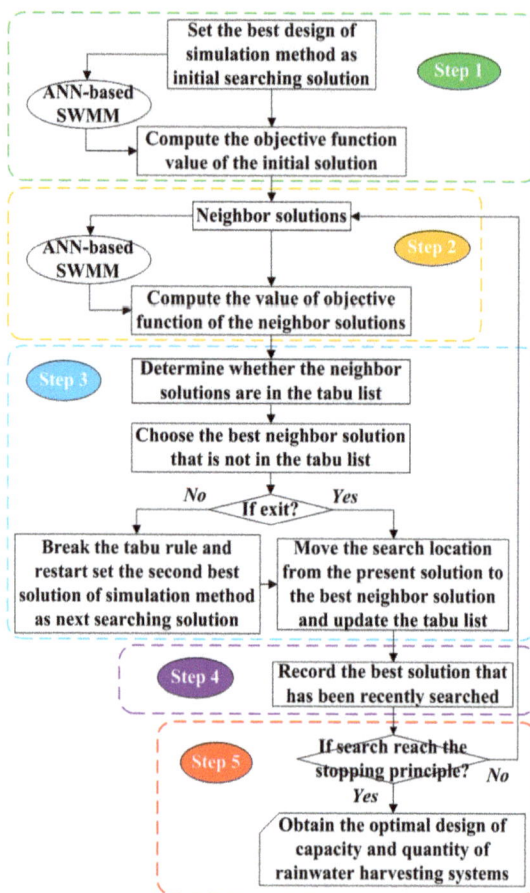

Figure 2. Flowchart of optimizing the spatial design of capacity and quantity of rainwater harvesting systems using tabu search

Step 1: Set the initial searching solution of tabu search (the best solution in the simulation method), input the alternative BPNN model, and calculate the objective function value of the optimization model; *i.e.*, the annual net benefit of the design pattern.

Step 2: Calculate the objective function value of the neighboring solution and choose the best neighboring solution.

Step 3: Check if the best neighboring solution is in the tabu list. If a best solution has already been searched, select the second best neighboring solution; if a best solution has not yet been searched, move the search location from the present solution to the best neighbor solution. After moving, update the tabu list.

Step 4: To record the optimal solutions identified thus far, apply the elite strategy to compare the best searching solution in this iteration with the optimal solution prior to the search.

Step 5: After the search principles stop working, the optimal spatial arrangement and capacity design approach for rain barrels can be obtained for the whole event.

2.5. Development of BPNN-Based SWMM

2.5.1. Model Structure of BPNN-based SWMM

This study develops a novel alternative BPNN-based simulation model to substitute for US EPA SWMM and embed it into the optimization model for the fast, accurate and automated optimizing process. Any newly considered interdisciplinary multi-objective optimization model, embedded simulation model and optimizing algorithm can be involved and integrated. Chang et al. (2010) [29] developed a two-stage procedure underlying the clustering-based hybrid inundation model, which is composed of linear regression models and ANNs (artificial neural networks) to build a 1-h-ahead regional flood inundation forecasting model. However, regarding to the study of modeling long lead-time continuous unsteady inundation level of an urban drainage system using ANNs still have not been researched. The inputs of BPNN-based alternative model include the boundary condition, initial condition and simulated target that must be entered in SWMM (e.g., precipitation, the LID design approach and water level of drainage system). Because the aim of this study is to determine flooding loss, the flooding and water pipe level are included in the input item. In addition, to obtain the best design approach for rain barrel spatial arrangement and capacity, we include the quantity and capacity of the rain barrels in the input item and set the water level/flooding at $t + 1$ moment as the output item. Besides, the success of BPNN-based simulation approach is mostly dependent on construction data (including training and validation part), which means data should be representative well enough in order to construct the input-output relation. In order to achieve this goal, this study develops a retrieving method of representative construction data using statistical quartiles analysis for rooftop area and rainfall frequency analysis (Section 2.2.2).

The BPNN-based drainage system water level simulation model is constructed in three parts: a single-moment training, single-moment validation, and complete-event simulation and verification. The single-moment simulation training and verification is primarily the calculation mechanism of the SWMM steady simulation. The complete-event simulation and verification adds the single-moment calculation units and provides feasibility verification for the unsteady simulation. In the complete-event simulation, we require a boundary condition at the start time (t_0), the meteorological-hydrological condition, and the spatial design pattern of the RWHS. We enter the trained BPNN single-moment calculation units and then obtain the simulation value at $t + 1$ from the output item. Accordingly, the cycle of continuous iterative calculations is repeated until the end of the moments, when the complete-event flooding and water levels of the drainage system can be simulated.

The BPNN was developed by Rosenblatt (1958) [19] and Rumelhart and McClelland (1986) [20]. Constructed by the multilayer perceptron, it belongs to a multilayer feedforward network and handles the nonlinear relationship between the input and output by a supervisual learning approach. The commonly used BPNN is a three-tier structure neural network, which includes an input layer, a hidden layer, and an output layer. The input value of the neurons connected by associated weights between different layers in the network is directly transferred into the hidden layer. Then, after the weighted accumulation (f), we obtain an output value and pass it onto the output layer following the same rule. The output value (y_j^n) of number j of the n-th layer is the conversion function value of the $n - 1$ layer neuron output value, which is shown as follows:

$$y_j^n = f(net_j^n) \tag{14}$$

The weight-accumulated value of the output value of the $n - 1$ layer net_j^n is shown as follows:

$$net_j^n = \sum_i w_{ji}^n y_i^{n-1} - b_j^{\ n} \tag{15}$$

In this study, the hidden layer adopts the tan-sigmoid (Equation (16)) as the transfer function, while the output layer is linear. BPNN utilizes the gradient steepest descent method to calculate and

31

adjust the network weight and bias values. This is accomplished to minimize the error of the output value and actual target value for obtaining a calculation mode of precise learning.

$$y_j = \frac{e^{net_j} - e^{-net_j}}{e^{net_j} + e^{-net_j}} \tag{16}$$

2.5.2. Alternative Applicability Assessing Index of BPNN-based SWMM

To assess if the developed BPNN-based SWMM is capable to be the alternative model of operating interface-restricted SWMM, this study adopts the mean absolute error (*MAE*) and coefficient of correlation (*CC*) as alternative applicability index, which are described below.

(1) *MAE*

$$MAE = \frac{\sum\limits_{i=1}^{n} \left| Y_{sim}^{BPNN}(t) - Y_{target}^{SWMM}(t) \right|}{n} \tag{17}$$

where $Y_{sim}^{BPNN}(t)$ is the simulation value of BPNN-based SWMM at time t, $Y_{target}^{SWMM}(t)$ is the target value to substitute US EPA SWMM, and n is the number of data. A smaller *MAE* indicates that the alternative applicability of the BPNN-based SWMM is better than the other BPNN-based models.

(2) *CC*

$$CC = \frac{n \sum Y_{sim}^{BPNN}(t) Y_{target}^{SWMM}(t) - \sum Y_{sim}^{BPNN}(t) \sum Y_{target}^{SWMM}(t)}{\sqrt{\sum \left(Y_{sim}^{BPNN}(t)\right)^2 - \frac{\left(\sum Y_{sim}^{BPNN}(t)\right)^2}{n}} \sqrt{\sum \left(Y_{target}^{SWMM}(t)\right)^2 - \frac{\left(\sum Y_{target}^{SWMM}(t)\right)^2}{n}}} \tag{18}$$

A larger *CC* indicates that the variation trend between the simulation value of BPNN-based SWMM and US EPA SWMM is closer that represents the developed BPNN-based SWMM is more suitable to be the alternative model of US EPA SWMM than the other BPNN-based models.

3. Application

3.1. Study Area

The Zhong-He District is an area of 20.29 km^2 located in the southwest corner of the Taipei Basin. Its southern end has a high altitude and gradually lowers northward. In some areas, the Zhong-He District has extreme slope changes, which can lead to floods because of the locations of these changes at the intersections of mountainous terrain and the ground. Other areas are also vulnerable to flooding on account of their more gentle terrains or insufficient drainage capacities. Examples include the area near Jyu-Guang Road and Min-Siang Street, shown in Figure 3a; control point 1 (CP1), Guo-Guang Street; control point 2 (CP2), Min-Siang Street; and control point 3 (CP3), Jyu-Guang Road. These latter three locations are low lying such that the terrain height diagram can be shown in Figure 3b. It is, therefore, relatively difficult for the water to drain from these areas, causing flooding and life and property loss from rainstorms. Thus, these locations are set as control points for the flood damage assessment.

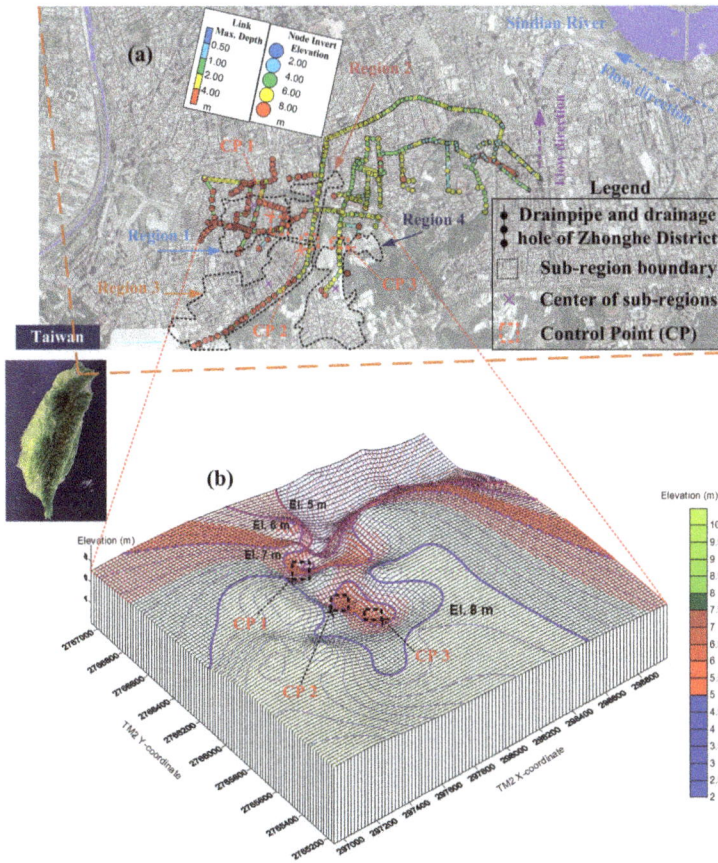

Figure 3. Study area: (**a**) spatial distribution of drainage system, zonal subregions for design of rainwater harvesting system using the fuzzy C-means cluster algorithm and the low-lying control points; (**b**) terrain height above sea level.

3.2. Analyzed Results of the Simulation Model for Spatial Design of Quantity and Capacity

3.2.1. Classified Results of Zonal Subregions for Design of RWHS

This study applies FCM cluster algorithm with practical investigated rooftop area data to classify the study area to characteristic zonal subregions. In order to choose a most economic mode of zonal subregions, this study precedes sensitivity analysis to different number of clusters for the distance of central locations. The distance of each two central locations for clustering central number 3 ranges from 547 m^2 to 722 m^2; distance for clustering number 4, ranges from 547 m^2 to 1075 m^2; distance for clustering number 5, ranges from 391 m^2 to 1117 m^2; and distance for clustering number 6, ranges from 375 m^2 to 1134 m^2. The average distance between each combination of two central locations for clustering number 3 to 6 are 656 m^2, 714 m^2, 673 m^2 and 685 m^2, respectively. Hence, clustering number 4 can cover wider designed area than the other clustering numbers with most efficient zonal mode.

After setting the clustering center number as 4 and calculating using FCM cluster algorithm, the catchment range of zonal subregions (Figure 3a). The four center coordinates (TM2 X, TM2 Y) are Region 1 (296565, 2765681), Region 2 (297360, 2764958), Region 3 (296860, 2765182) and Region 4 (297247, 2765702), respectively. The number of available building roof for arranging rain barrel of

Region 1 is 440 which is mostly composed of schools and residences; number of available roof of Region 2 is 385, composed of parks and commercial buildings; number of Region 3 is 943, composed of community high buildings and housing; and number of Region 4 is 728, composed of industrial buildings and residences.

3.2.2. Spatial Designed Results of Specific Representative Regular RWHS

(1) Capacity and Quantity

The designed parameters for RWHS include capacity (volume) and quantity (arranged density: how many areas (A_l) arrange one rain barrel). This study designs representative regular specification of RWHS by using statistical quartiles analysis (to estimate representative A_l) and rainfall frequency analysis (to estimate representative $P_T^{\hat{R}P}$). The statistical quartiles analysis results of available rooftop area of each subregion on Zhong-He drainage area is shown in Figure 4. This study adopts $Min(a_r^{min})$, $WA(a_r^{min})$, $WA(a_r^{25\%})$ and $WA(a_r^{med})$ among four subregions ($r = 1$–4) that the values are 55.0 m^2 (A_1), 82.4 m^2 (A_2), 108.5 m^2 (A_3) and 152.0 m^2 (A_4), respectively, to ensure all designs of volume and arranged density can actually be applied to the building of Zhong-He drainage area. The adopted return period (T) of $P_T^{\hat{R}P}$ are 2, 5, 25, 50 and 100 years, and the designed rainfall duration is 6 hours. Finally, the designed regular volume of rain barrel are $A_1 \cdot P_{2year}^{\hat{R}P}$, $A_2 \cdot P_{5year}^{\hat{R}P}$, $A_2 \cdot P_{50year}^{\hat{R}P}$, $A_3 \cdot P_{50year}^{\hat{R}P}$, $A_4 \cdot P_{25year}^{\hat{R}P}$ and $A_4 \cdot P_{100year}^{\hat{R}P}$ that the values are 3.03 m^3 (S_1), 6.14 m^3 (S_2), 9.12 m^3 (S_3), 12.01 m^3 (S_4), 15.05 m^3 (S_5) and 18.05 m^3 (S_6), respectively, to ensure that the designed volume can handle all kinds magnitude of storm rainwater of return periods.

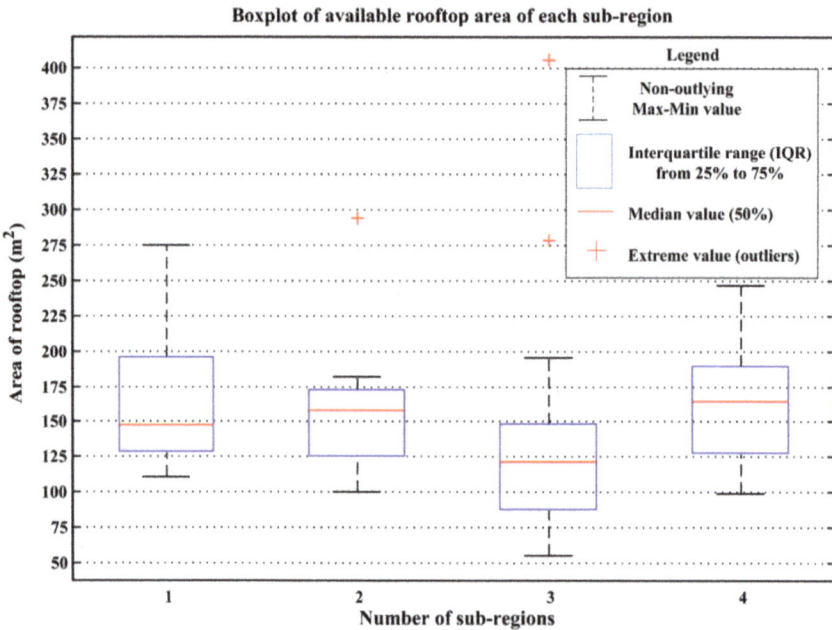

Figure 4. Boxplot of available rooftop area of each subregion in the study area.

(2) Spatial Arrangement of Designed Cases

There were 284 sub-catchments in the study area. Areas with rain barrels included business, mixed residential, industrial, office, and school districts. We conducted designs of the space, density, and capacity to set the locations of the rain barrels. According to the spatial design style, we established four types of arrangements (Cases 1–4) for the simulation method. The rain barrels of Case 1 were arranged at whole sub-catchments; those for Case 2 were set mainly at inundated sub-catchments; those for Case 3 were arranged at easily inundated sub-catchments but without outer space; and those for Case 4 were set upstream from the inundated sub-catchments. Figure 5 shows the urban drainage system setup for the three Zhong-He District cases.

Figure 5. Spatial arrangement of regular design of rainwater harvesting systems.

In terms of the setup of the rain barrel quantity, we employed density to establish it in SWMM. To compare its flood detention effects, we set the density as: (1) one for every 55.0 m^2 under the rain barrel per household (Case X-1); one for every 82.4 m^2 (Case X-2); one for every 108.5 m^2 (Case X-3); and one for every 152.0 m^2 (Case X-4). The rain barrel quantity of each case with each spatial arrangement is shown in Table 1. In the capacity design, we divided the capacity of rain barrels into 3.03 m^3 (Case X-Y-1), 6.14 m^3 (Case X-Y-2), 9.12 m^3 (Case X-Y-3), 12.01 m^3 (Case X-Y-4), 15.05 m^3 (Case X-Y-5), and 18.05 m^3 (Case X-Y-6) to compare the simulated effect of flood detention.

Table 1. Quantity (number) of rain barrel of each designed regular case.

Case No.	Case X-1	Case X-2	Case X-3	Case X-4
Case 1	3194		X	
Case 2	472	317	236	167
Case 3	306	204	153	108
Case 4	395	262	198	136

3.2.3. Calibration and Validation of US EPA SWMM

The distribution of sewer system construction and flood control point are shown in Figure 3a. Because the New Taipei sewer water level and flow monitoring system had not yet been built, the model calibration and validation could only be executed within the range of flooding depth. We therefore employed the flooding areas, water logging time, receding time, and flooding depth from "12 August 2009 Rainstorm Survey Data" for the model calibration. In addition, we used the "16 June 2012 Rainstorm Event" for the model validation to examine its feasibility.

The simulated output of SWMM was the volume of burst pipes (flooding) and water level; therefore, the flooding had to be converted into flooding depth for comparison. The actual records of the calibrated event's three control points and SWMM simulation result both showed flooding; however, the validated event's actual record and SWMM simulation result showed that only CP3 had flooding, whereas no flooding was found at the other control points. We checked the simulated calculation results of the calibration and validation event. The simulated depths of each flooding control point were all located within the actual flooding record range (Figures 6 and 7). It was therefore confirmed that the model parameters were well calibrated and complete. The values of calibrated parameters are shown in Table 2.

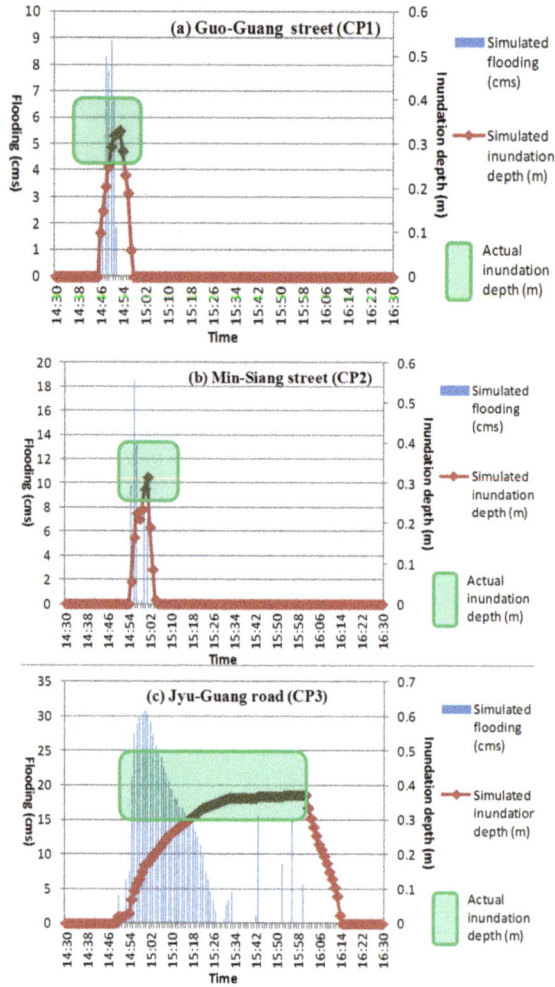

Figure 6. Calibration results of stormwater runoff management model (SWMM).

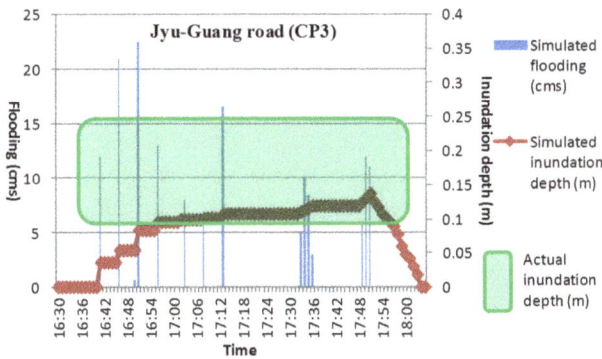

Figure 7. Validation results of SWMM (Jyu-Guang Road).

Table 2. Values of the calibrated parameters in SWMM.

Title	Surface Roughness Coefficient	Horton-Based Max/Min Infiltration Rates/Decay Constant	Manning's Roughness Coefficient for Conduits
Region 1	0.015~0.033	3.5~4.6 (mm/h)/0.9~1.6 (mm/h)/1.8~2 (1/h)	0.013~0.016
Region 2	0.013~0.037	3.6~5.8 (mm/h)/1.1~1.9 (mm/h)/1.6~1.9 (1/h)	0.015~0.017
Region 3	0.013~0.025	3~3.5 (mm/h)/0.5~0.9 (mm/h)/2~2.2 (1/h)	0.011~0.015
Region 4	0.012~0.021	3.3~3.9 (mm/h)/0.7~1.2 (mm/h)/1.9~2 (1/h)	0.012~0.014

3.2.4. Simulated Analytical Results of the Designed Spatial Arranged Regular Cases for RWHS

(1) Rainstorm Events Employed for Simulation-Optimization

As the foundation for selecting the optimal spatial design of RWHS, we gathered data from four rainstorm events that occurred from 2009 to 2012 in the Zhong-He District which the representative return period of total precipitation are 50, 100, 125 and 75 year, respectively (Table 3). The duration of heavy storm rains were about 6 hours which all caused large amount of inundation loss.

Table 3. Adopted precipitation hyetograph of rainstorm events for simulation-optimization (mm).

Date	Time Series Number (HOUR)						Representative Return Period of Total Precipitation
	1	2	3	4	5	6	
1 July 2009	4.0	96.5	11.0	3.0	2.5	0	50 year
12 August 2009	2.5	101.5	20.0	3.5	1.5	0	100 year
21 June 2010	4.0	0	4.5	44.5	48.0	28.5	125 year
12 August 2012	70.5	49.5	1.5	0	0	0	75 year

(2) Results and Discussion

We entered the spatial design approach of all cases into SWMM to simulate and calculate the average annual net benefit of setting the RWHS in the rainstorm events within the study years. We then compared the result to a scenario without RWHS. The curve equations of inundation loss of CP1–CP3 are expressed in Equations (19)–(21), respectively, and the annual cost function of RWHS is expressed in Equation (22). The unit of L_{CP1}, L_{CP2}, L_{CP3} and T_C is US dollars, and the unit of F_{CP1}^{total}, F_{CP2}^{total} and F_{CP3}^{total} is m^3. The unit inundation loss of CP_2 is obviously larger than the other two control points.

$$L_{CP1}=3361.60 \cdot \left(F_{CP1}^{total} \right) + 25845.03 \quad R^2 = 0.9565 \tag{19}$$

$$L_{CP2}= -8.89 \cdot \left(F_{CP2}^{total} \right)^2 + 8564.43 \cdot \left(F_{CP2}^{total} \right) - 9.67 \cdot 10^4 \quad R^2 = 0.9719 \tag{20}$$

$$L_{CP1} = 1.42 \times 10^5 \cdot \left(F_{CP1}^{total} \right)^{0.3681} \quad R^2 = 0.9709 \tag{21}$$

$$T_C = 118.21 + 4.34 \cdot C_a^2 + 3.32 \cdot A_r \tag{22}$$

The results of Case 1 can be regarded as the most significant flood reduction effect for the RWHS. However, because the cost of RWHS was very large, all net benefits resulted in a negative value. The design approach for the largest net benefit in the Cases 2, 3 and 4 considered individually are Cases 2-1, 3-1, and 4-1, respectively, and the comparison varying along with volumes is shown in Figure 8. The simulation analysis results demonstrate: (1) The function of flooding damage and reserving the facilities cost for each spatial layout appeared as convex and concave curves, thereby changing with the capacity of the rain barrels. We subtracted the convex curve from the concave curve to obtain the best solution with the largest net benefit; (2) The best solution was when the RWHS were set upstream of the flooding area, which was Case 4-1; the capacity of the rain barrel was 12 m^3, and the net benefit for each year was 4.61 × 10^5 US dollars; (3) In each case, the rain barrel's best capacity was between 12 and 15 m^3; greater benefits were produced when the rain barrel was set in the easily flooded area.

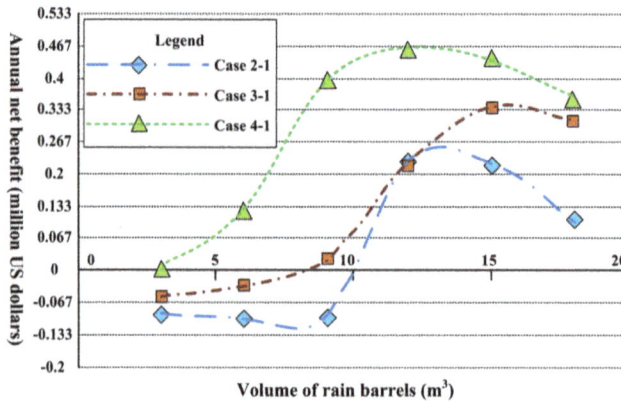

Figure 8. Comparative diagram of annual net benefit of each regularly designed case.

3.3. Construction Results of BPNN-based SWMM

There were 12 input items of BPNN-based SWMM, which included: catchment precipitation, CP1−CP3 full pipe percentage of water flow, the quantity of Regions 1−4 rain barrels, and the capacity of Regions 1−4 rain barrels. The training and validating results are described as follows:

3.3.1. Training and Validating Results of Single-moment Simulation

We used the 2:1 principle to divide the rainfall-runoff data of the urban drainage system generated in each event using the simulation methods into two parts—training and validation. When classifying, we strove to disperse them in cases with different designs. The quantity of all events was 288; *i.e.*, 3 (spatial arrangement quantity) × 4 (rain barrel intensity quantity) × 4 (storm event quantity) × 6 (rain barrel capacity). There were 198 training events and 90 validation events. Because the time interval for the data of each event was 1 min, the sum of training data was 44,442 and the sum of the validation data was 20,070.

After several repeated tests of neurons in the hidden layer, we compared the appraisal indicators and determined that there were seven final hidden layer neurons. The simulated results of training and validation for full pipe percentage of water flow are shown in Figures 9 and 10, respectively. The *MAE* of CP1−CP3 in a single-moment level full-pipe simulation among validation events was 0.010%, 0.014% and 0.032%, respectively. The *CC* value was 0.990, 0.995, and 0.983, respectively. The results showed that the single-moment error of the full pipe percentage was small and demonstrated an accurate simulation trend. Therefore, the training and validation results of the single-moment simulation were good and could be continued in the entire-event simulation. However, the *MAE* of CP1−CP3 in a single-moment volume burst-pipes simulation among validation events was 0.020 cms, 0.039 cms, and 0.808 cms, respectively. The *CC* value was 0.947, 0.783, and 0.916, respectively. The results indicated that BPNN-based SWMM demonstrated better performance for water level simulation. Nevertheless, it was more difficult to be accurate in terms of the volume of burst pipes.

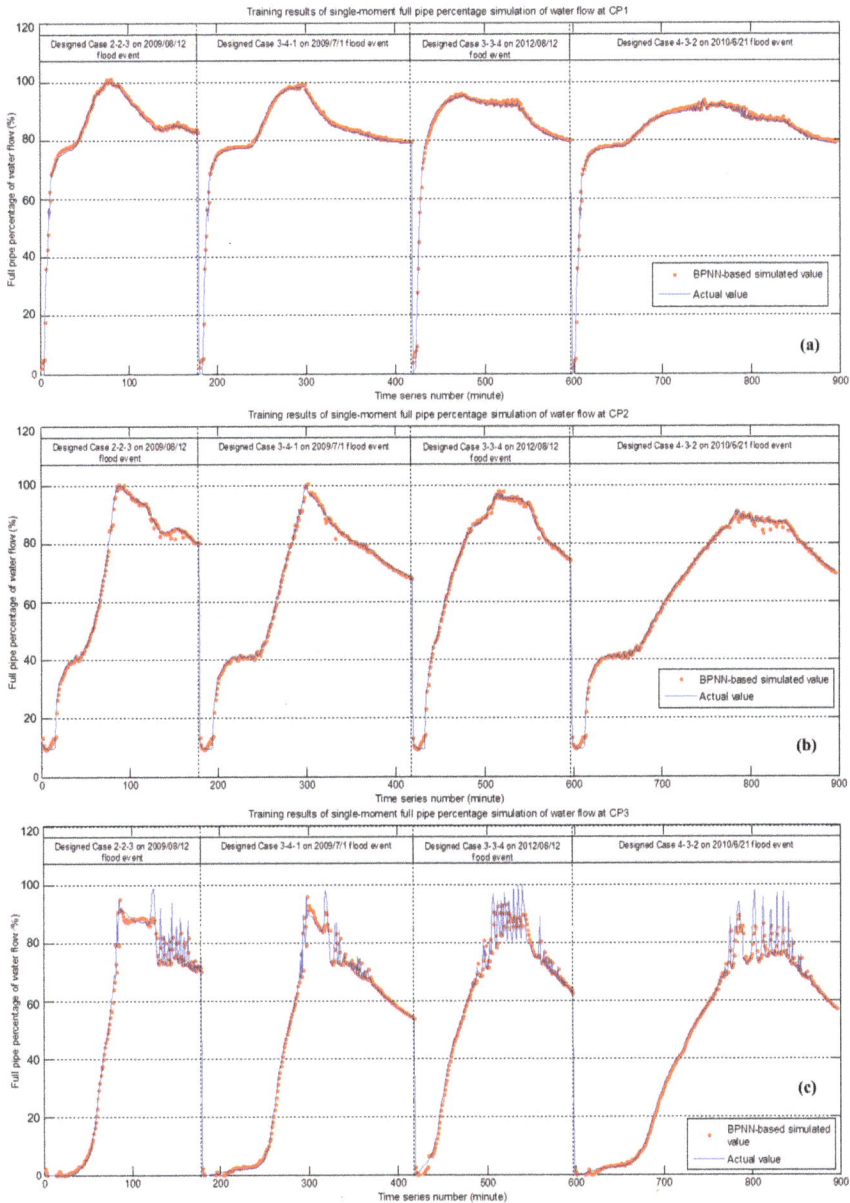

Figure 9. Training results of single-moment full pipe percentage simulation of water flow: (**a**) at CP1; (**b**) at CP2; and (**c**) at CP3.

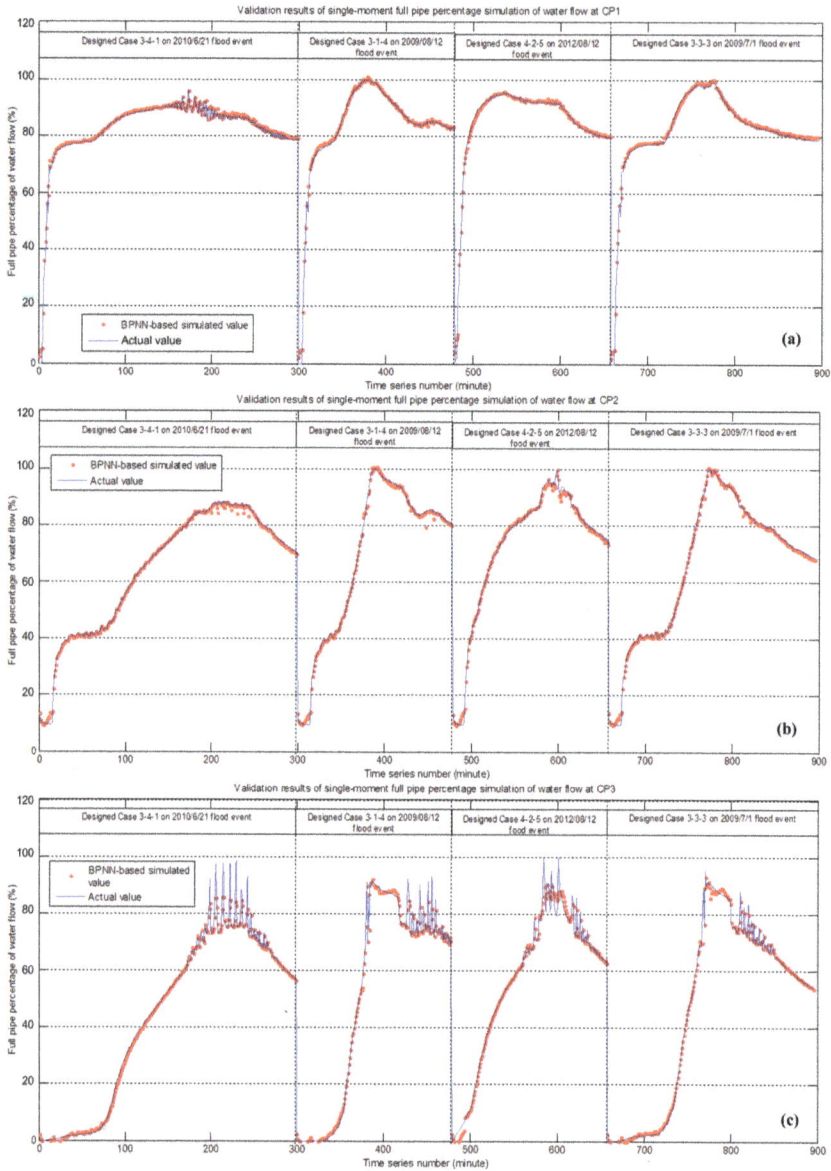

Figure 10. Validation results of single-moment full pipe percentage simulation of water flow: (**a**) at CP1; (**b**) at CP2; and (**c**) at CP3.

3.3.2. Sensitivity Analysis Result

To understand the performance of the proposed BPNN-based water flow simulation model, this study deeply performed sensitivity analysis and the results are shown in Table 4. Results show that the output (full pipe percentage at time $t + 1$) sensitivity of CP3, CP2 and CP1 with regard to input: precipitation at time t (change in 0.2 mm/min) is 3.46%, 5.15% and 5.26%, respectively. According to historical experimental records, assuming about 67% of flood can be removed by free flow outlet and

40

pumping facilities, the drainage system can suffer about 12.1 mm/h of heavy rains with no flooding that coincide with the rainfall design standard of 5-year return period (12.4 mm/h), so it represents the model performance and capability for the input of precipitation is available. Furthermore, the output sensitivity of CP1−CP3 with regard to input: full pipe percentage at *t* (change in 1%/min) is within the range from 0.51% to 1.12%. The change in the downstream water flow of CP1 is more sensitive to the other control points that coincide with the hydraulic theory, so the developed model is scientific enough to model water flow phenomenon. Besides, the output sensitivity of CP1−CP3 with regard to input: arranged quantity (change in 100 numbers) is within the range from 1.81% to 6.75%, and the output sensitivity of CP1−CP3 with regard to input: arranged capacity (change in 3 m³) is within the range from 0.44% to 2.15%. The change in upstream quantity and capacity of RWHS (Regions 3 and 4) make more sensitivity to the other low-lying subregions (Regions 1 and 2) that coincide with the analytical results of simulation method (Section 3.2.4), so the developed model is available for the embedded optimizing process.

Table 4. Sensitivity analysis of the backpropagation neural network (BPNN)-based water flow simulation model.

Input	Average Variance of Output (Full Pipe Percentage)		
	of CP3 at $t + 1$	of CP2 at $t + 1$	of CP1 at $t + 1$
Precipitation at time t (change in 0.2 mm/min)	3.46%	5.15%	5.26%
Full pipe percentage of CP3 at t (change in 1%/min)	1.12%	0.51%	0.72%
Full pipe percentage of CP2 at t (change in 1%/min)	0.59%	1.00%	0.53%
Full pipe percentage of CP1 at t (change in 1%/min)	0.96%	0.98%	1.00%
Arranged quantity of Region 1 (change in 100 number)	3.22%	2.68%	2.74%
Arranged quantity of Region 2 (change in 100 number)	2.38%	3.54%	1.81%
Arranged quantity of Region 3 (change in 100 number)	2.40%	5.02%	3.42%
Arranged quantity of Region 4 (change in 100 number)	4.40%	3.30%	6.75%
Arranged capacity of Region 1 (change in 3 m³)	1.43%	1.19%	0.88%
Arranged capacity of Region 2 (change in 3 m³)	0.49%	0.86%	0.44%
Arranged capacity of Region 3 (change in 3 m³)	1.43%	2.15%	1.75%
Arranged capacity of Region 4 (change in 3 m³)	1.16%	1.29%	1.62%

3.3.3. Validation Results of the Entire-event Iterative Simulation

The validation results of the entire-event iterative continuous simulation of the developed BPNN-based model are shown in Table 5. The *MAE* of the water level simulation was quite small (less than 15% for all cases), and the *MAE* for CP1−CP3 was 0.065%, 0.07%, and 0.106%, respectively. Moreover, all *CC* values reached 0.96; the CC values for CP1−CP3 were 0.968, 0.970, and 0.963, respectively. These results indicate that the BPNN-based SWMM developed by our research can accurately and quickly simulate the water level change of rainstorm events. Therefore, its alternative model can be reliable embedded in the optimization model to quickly and automatically provide an optimal design approach while the window-based man-made operating interface of hydraulic model cannot be linked with optimization model and algorithm. Figure 11 depicts the validation result of Case 4-3-6 during the "12 August 2012 Rainstorm Event", which represents the fourth spatial distribution, the third rain barrel arrangement intensity (one for every 150 m²), and the second design capacity (6 m³).

Table 5. Unsteady continuous simulated results for validation of BPNN-based SWMM.

Unsteady Simulated Events	Guo-Guang Street (CP1)		Min-Xiang Street (CP2)		Ju-Guang Road (CP3)	
	MAE (%)	*CC*	*MAE (%)*	*CC*	*MAE (%)*	*CC*
Designed Case 3-1-3 on 1 July 2009	5.7	0.974	7.0	0.975	9.2	0.974
Designed Case 4-4-1 on 1 July 2009	5.2	0.984	7.5	0.981	9.6	0.975
Designed Case 2-1-2 on 12 August 2009	3.8	0.995	3.0	0.994	10.1	0.963
Designed Case 3-2-4 on 12 August 2009	4.1	0.995	3.6	0.994	9.4	0.973
Designed Case 3-3-1 on 21 June 2010	12.4	0.974	10.1	0.991	17.1	0.936
Designed Case 4-3-6 on 21 June 2010	11.1	0.970	7.0	0.991	15.8	0.956
Designed Case 2-2-5 on 12 August 2012	5.1	0.951	6.9	0.951	7.2	0.967
Designed Case 4-1-6 on 12 August 2012	5.9	0.924	9.5	0.914	9.8	0.954
Designed Case 4-3-2 on 12 August 2012	5.5	0.941	7.9	0.940	7.0	0.964
Average	6.5	0.968	7.0	0.970	10.6	0.963

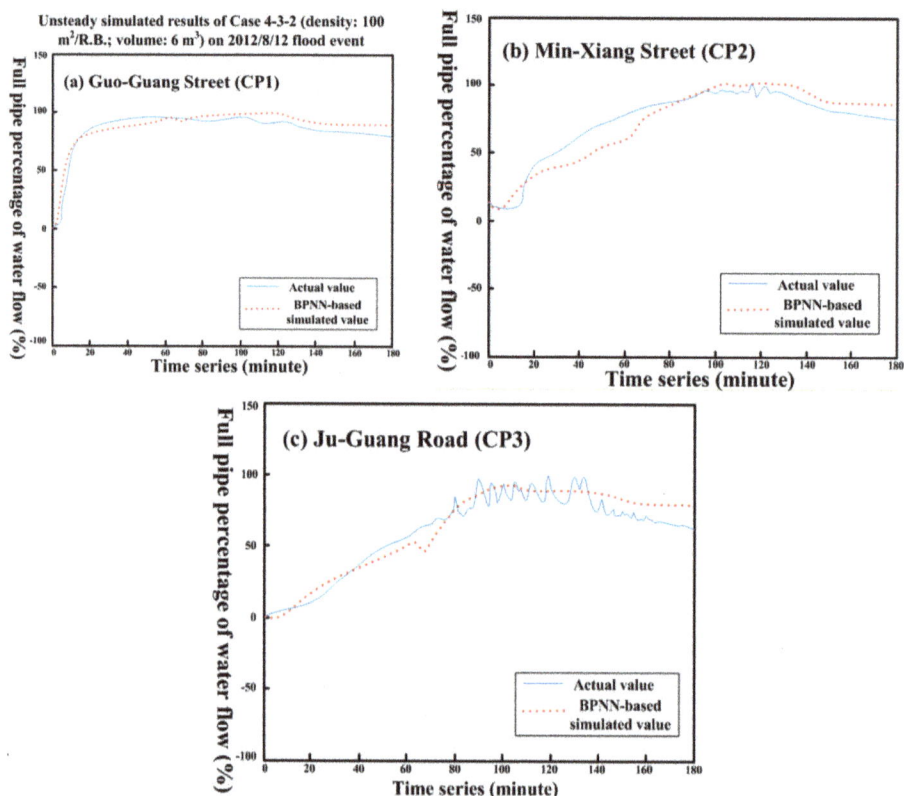

Figure 11. Unsteady continuous simulated results of Case 4 during the "12 August 2012 Flood Event" using BPNN-based SWMM.

3.4. Optimization Results

In this applied case, the tabu list can be shown as $[Z, N_1, N_2, N_3, N_4, S_1, S_2, S_3, S_4]$. Moreover, the tabu list length was set to 300, and the searching iterative number was 1,000. In the searching solution, the moving distance of S_r and N_r was 1 and 10, respectively. The initial solution was the best design approach of the simulation method of Case 4-1-4 (one rain barrel for every 50 m²; capacity of 12 m³).

After optimizing by tabu search, the optimal searching results are shown in Figure 12, and comparison on full pipe percentage of water flow and flooding volume between optimal rain barrel design, best design of simulation method and original no design of the three control points is shown in Figure 13. Results show that because of arrangement of rain barrels, the optimized design and best design of simulation method can eliminate total flood in the drainage system as much as possible comparing to original circumstances of no rain barrel. However, because the best design of simulation method mostly arrange rain barrels on the upstream of the study area, the initial stored effect for flood mitigation is obvious. However, after the rain barrels are full, the overflow from upstream would impact the low-lying stored water sharply with large amount of momentum because of highly hydraulic gradient. These circumstances cause the low-lying control points would instead undertake flooding transitorily. Moreover, the optimized design can eliminate the peak flow as much as possible and adapt the flow velocity as less as possible to minimize the flooding at the control points, because of more elastic spatial arrangement considering the distribution of drainage system and terrain. In the optimal spatial design approach for rain barrels, spatial quantity was mainly located upstream, and rain barrels with greater volume in easily flooded areas had a better flood reduction effect.

The BPNN-based SWMM developed by our institute may have some errors in the calculation results; therefore, our institute returned the optimal solution to the US EPA SWMM to simulate the water level and volume of the burst pipes, and to calculate the actual flooding loss. Results indicate that the evaluated error of inundation loss by using BPNN-based SWMM comparing to US EPA SWMM is about 7.99%; and regarding the evaluated error of net benefit, is about 4.15%; that is within acceptable range. The stepwise calculated results for net benefit are shown in Table 6. The average inundated loss while no installing rain barrels was 1.04×10^6 US dollars; and of optimized design, was 0.27×10^6 US dollars. The optimized spatial design of RWHS could reduce 72% of inundation losses according to the four simulated flood events. Besides, the annual net benefit of the best solution in the simulation method was 4.61×10^5 US dollars (Figure 13), and the annual net benefit of hybrid simulation-optimization method was 5.20×10^5 US dollars (12.75% better than using the single simulation method), which is quite good. It indicates that the optimization model developed by our institute can search for the optimal solutions for spatial quantity and capacity arrangement of RWHS with consideration of flood retention benefits.

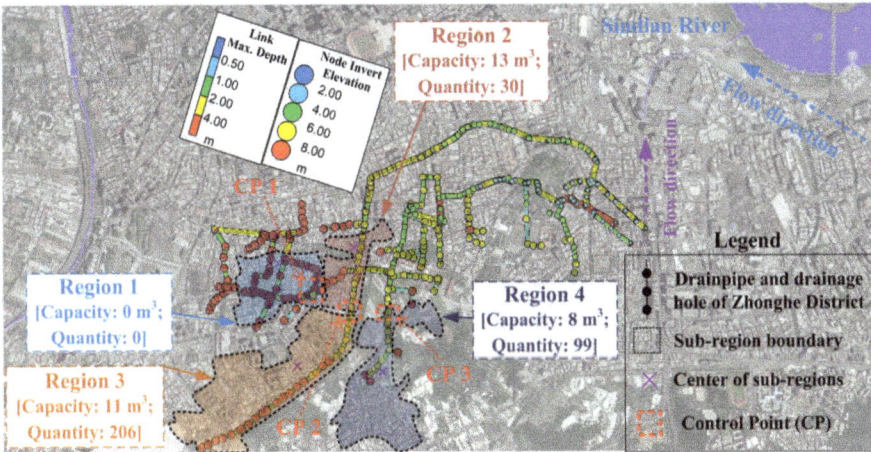

Figure 12. Optimized design results of capacity and quantity of rainwater harvesting systems for Zhong-He drainage system.

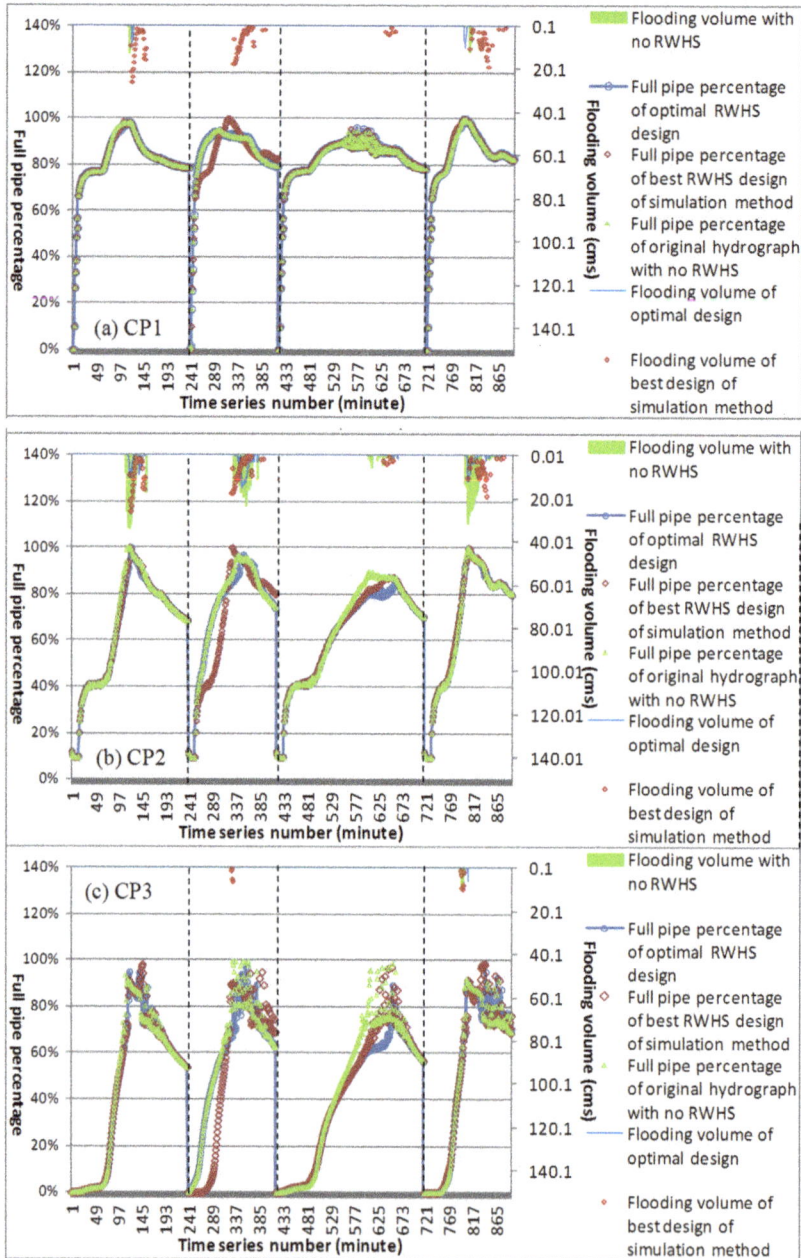

Figure 13. Comparison on full pipe percentage of water flow and flooding volume between optimal rainwater harvesting systems (RWHS) design, best design of simulation method and original no design: (a) CP1; (b) CP2; and (c) CP3.

Table 6. Net benefit of optimized design of rainwater harvesting systems.

Flood Event	1 July 2009	12 August 2009	21 June 2010	12 August 2012	Average
Inundated loss while no installing rain barrels (US dollars)	1.03×10^6	1.63×10^6	0.58×10^6	0.91×10^6	1.04×10^6
Inundated loss of optimized design (US dollars)	0.25×10^6	0.33×10^6	0.22×10^6	0.27×10^6	0.27×10^6
Decreased inundated loss (US dollars)	0.79×10^6	1.30×10^6	0.36×10^6	0.64×10^6	0.77×10^6
Benefit percentage of flood mitigation (%)	76.2%	79.8%	62.1%	70.0%	72.0%
Cost (US dollars)			0.25×10^6		
Annual net benefit (US dollars)			0.52×10^6		

4. Conclusions

This study established a set of simulation-optimization models to select optimal solutions for spatial quantity and capacity arrangements of rain barrels in urban drainage areas. These models consider the optimal net benefit and flooding loss/cost reduction. First, we classified the characteristic zonal subregions for design of rainwater harvesting system (RWHS) by using FCM cluster algorithm with the investigated data of urban roof, land use and rainfall characteristics of drainage area, and the representative regular specification of spatial quantity and capacity arrangement of different types of RWHS are designed by using statistical quartiles analysis for rooftop area and rainfall frequency analysis. In the simulation method, we used SWMM built with actual rainfall data to simulate the water level and volume of burst pipes for a drainage system, and to obtain various flood reduction situations from different designs of regular RWHS for spatial quantity and capacity arrangements. We then calculated the net benefit after deducting the cost. In the optimization method, we first established a water level simulation model that can substitute for US EPA SWMM based on the simulated analysis results combined with BPNN, and then embedded it into the optimization model. Finally, considering the flood reduction benefit, we combined the optimization model and tabu search to optimize the spatial design approach for the quantity and capacity arrangement of RWHS.

This study applied the established approach to Zhong-He District in New Taipei City, Taiwan. The research method is innovative and combines various forms of artificial intelligence as well as methods and techniques of systems analysis to effectively and fast optimize design approach of RWHS. Results showed that the developed hybrid simulation-optimization approach which embedded an intelligent BPNN-based water flow simulation model and used tabu search to obtain the optimal solution of the optimization model was 12.75% better than using the single simulation method comparing to economic net benefit for flood mitigation. The optimized spatial design of RWHS could reduce 72% of inundation losses according to the four simulated flood events. Furthermore, the optimal spatial design approach for rain barrels indicate that spatial quantity was mainly located at upstream and meanwhile RWHS with greater volume in easily flooded areas had a better flood reduction effect. Besides, the developed embedded BPNN-based SWMM for unsteady continuous water level simulation of drainage system can achieve average 92% of accuracy in the three control points. This capability promotes the developed simulation-optimization procedure can: (1) quickly and effectively search for the optimal solution; (2) conform to newly considered interdisciplinary multi-objective/constraints; and (3) involve more related embedded models. Moreover, the simulation-optimization process developed in this study can select a flexible and practical spatial arrangement and capacity design approach for RWHS to be an alternative measure for urban flood mitigation.

Acknowledgments: This research was partially supported by the Ministry of Science and Technology, Taiwan (Grant Nos. MOST103-2221-E-002-246 and MOST104-2111-M-019-001). In addition, the authors are indebted to the reviewers for their valuable comments and suggestions.

Author Contributions: Chien-Lin Huang performed the model construction and experiments, analyzed the data and wrote the paper; Nien-Sheng Hsu and Chih-Chiang Wei conceived and designed the models and experiments. Wei-Jiun Luo performed the model construction and experiments.

Conflicts of Interest: The authors declare no conflict of interest.

References

1. Liaw, C.H.; Tsai, Y.L. Optimum storage volume of rooftop rain water harvesting systems for domestic use. *J. Am. Water Resour. Assoc.* **2004**, *40*, 901–912. [CrossRef]
2. Liaw, C.-H.; Chiang, Y.-C. Framework for assessing the rainwater harvesting potential of residential buildings at a national level as an alternative water resource for domestic water supply in Taiwan. *Water* **2014**, *6*, 3224–3246. [CrossRef]
3. Chiu, Y.R.; Liaw, C.H.; Chen, L.C. Optimizing rainwater harvesting systems as an innovative approach to saving energy in hilly communities. *Renew. Energy* **2009**, *34*, 492–498. [CrossRef]
4. Campisano, A.; Modica, C. Optimal sizing of storage tanks for domestic rainwater harvesting in Sicily. *Resour. Conserv. Recycl.* **2012**, *63*, 9–16. [CrossRef]
5. Abdulla, F.A.; Al-Shareef, A.W. Roof rainwater harvesting systems for household water supply in Jordan. *Desalination* **2009**, *243*, 195–207. [CrossRef]
6. Belmeziti, A.; Coutard, O.; de Gouvello, B. A new methodology for evaluating potential for potable water savings (PPWS) by using rainwater harvesting at the urban level: The case of the municipality of Colombes (Paris Region). *Water* **2013**, *5*, 312–326. [CrossRef]
7. Aladenola, O.O.; Adeboye, O.B. Assessing the potential for rainwater harvesting. *Water Resour. Manag.* **2010**, *24*, 2129–2137. [CrossRef]
8. Hajani, E.; Rahman, A. Reliability and cost analysis of a rainwater harvesting system in peri-urban regions of greater Sydney, Australia. *Water* **2014**, *6*, 945–960. [CrossRef]
9. Pachpute, J.; Tumbo, S.; Sally, H.; Mul, M. Sustainability of rainwater harvesting systems in rural catchment of Sub-Saharan Africa. *Water Resour. Manag.* **2009**, *23*, 2815–2839. [CrossRef]
10. Seo, Y.; Park, S.Y.; Kim, Y.-O. Potential benefits from sharing rainwater storages depending on characteristics in demand. *Water* **2015**, *7*, 1013–1029. [CrossRef]
11. Su, M.D.; Lin, C.H.; Chang, L.F.; Kang, J.L.; Lin, M.C. A probabilistic approach to rainwater harvesting systems design and evaluation. *Resour. Conser. Recycl.* **2009**, *53*, 393–399. [CrossRef]
12. Baguma, D.; Loiskandl, W.; Jung, H. Water management, rainwater harvesting and predictive variables in rural households. *Water Resour. Manag.* **2010**, *24*, 3333–3348. [CrossRef]
13. Jones, M.P.; Hunt, W.F. Performance of rainwater harvesting systems in the southeastern United States. *Resour. Conserv. Recycl.* **2010**, *54*, 623–629. [CrossRef]
14. Basinger, M.; Montalto, F.; Lall, U. A rainwater harvesting system reliability model based on nonparametric stochastic rainfall generator. *J. Hydrol.* **2010**, *392*, 105–118. [CrossRef]
15. Palla, A.; Gnecco, I.; Lanza, L.G. Non-dimensional design parameters and performance assessment of rainwater harvesting systems. *J. Hydrol.* **2011**, *401*, 65–76. [CrossRef]
16. Burns, M.J.; Fletcher, T.D.; Duncan, H.P.; Hatt, B.E.; Ladson, A.R.; Walsh, C.J. The performance of rainwater tanks for stormwater retention and water supply at the household scale: An empirical study. *Hydrol. Process.* **2015**, *29*, 152–160. [CrossRef]
17. Campisano, A.; Modica, C. Appropriate resolution timescale to evaluate water saving and retention potential of rainwater harvesting for toilet flushing in single houses. *J. Hydroinform.* **2015**, *17*, 331–346. [CrossRef]
18. Petrucci, G.; Deroubaix, J.F.; de Gouvello, B.; Deutsch, J.C.; Bompard, P.; Tassin, B. Rainwater harvesting to control stormwater runoff in suburban areas, an experimental case study. *Urban Water J.* **2012**, *9*, 45–55. [CrossRef]
19. Rosenblatt, F. The perceptron: A probabilistic model for information storage and organization in the brain. *Psychol. Rev.* **1958**, *65*, 386–408. [CrossRef] [PubMed]
20. Rumelhart, D.E.; McClelland, J.L. *Parallel Distributed Processing: Explorations in the Microstructure of Cognition*; MIT Press: Cambridge, MA, USA, 1986.
21. Roesner, L.A.; Dickinson, R.E.; Aldrich, J.A. *Storm Water Management Model Version 4: User's Manual*; United States Environmental Protection Agency: Washington, DC, USA, 1988.

22. Rossman, L.A. *Storm-Water Management Model Version 5.0*; United States Environmental Protection Agency: Washington, DC, USA, 2005.

23. Glover, F. Future paths for integer programming and links to artificial intelligence. *Comput. Oper. Res.* **1986**, *13*, 533–549. [CrossRef]

24. Glover, F.; Laguna, M. *Tabu Search*; Kluwer Academic: Boston, TX, USA, 1997.

25. Cheng, C.L.; Liao, M.C. Regional rainfall level zoning for rainwater harvesting systems in northern Taiwan. *Resour. Conserv. Recycl.* **2009**, *53*, 421–428. [CrossRef]

26. Bezdek, J.C. *Pattern Recognition with Fuzzy Objective Function Algorithms*; Plenum Press: New York, NY, USA, 1981.

27. Lee, K.; Ho, J. Design hyetograph for typhoon rainstorms in Taiwan. *J. Hydrol. Eng.* **2008**, *13*, 647–651. [CrossRef]

28. Guo, J.; Blackler, G.; Earles, T.; MacKenzie, K. Incentive index developed to evaluate storm-water low-impact designs. *J. Environ. Eng.* **2010**, *136*, 1341–1346. [CrossRef]

29. Chang, L.C.; Shen, H.Y.; Wang, Y.F.; Huang, J.Y.; Lin, Y.T. Clustering-based hybrid inundation model for forecasting flood inundation depths. *J. Hydrol.* **2010**, *385*, 257–268. [CrossRef]

water

MDPI

Article

Designing Rainwater Harvesting Systems Cost-Effectively in a Urban Water-Energy Saving Scheme by Using a GIS-Simulation Based Design System

Yie-Ru Chiu [1], Yao-Lung Tsai [2],* and Yun-Chih Chiang [3]

[1] Center for General Education, Tzu-Chi University, No.701, Zhongyang Rd., Sec.3, Hualien 97004, Taiwan; chiuyr@gmial.com
[2] Department of Leisure and Recreation Studies, Aletheia University, No.70-11 Pei-Shi Liao, Matou, Tainan 72147, Taiwan
[3] Center for General Education, Tzu-Chi University, No.701, Zhongyang Rd., Sec.3, Hualien 97004, Taiwan; ycchiang@mail.tcu.edu.tw
* Author to whom correspondence should be addressed; yltsai2011@gmail.com;
   Tel.: +886-6-570-3100 (ext. 7419); Fax: +886-6-570-3834.

Academic Editor: Ataur Rahman
Received: 20 September 2015; Accepted: 3 November 2015; Published: 10 November 2015

Abstract: Current centralized urban water supply depends largely on energy consumption, creating critical water-energy challenge especially for many rapid growing Asian cities. In this context, harvesting rooftop rainwater for non-potable use has enormous potential to ease the worsening water-energy issue. For this, we propose a geographic information system (GIS)-simulation-based design system (GSBDS) to explore how rainwater harvesting systems (RWHSs) can be systematically and cost-effectively designed as an innovative water-energy conservation scheme on a city scale. This GSBDS integrated a rainfall data base, water balance model, spatial technologies, energy-saving investigation, and economic feasibility analysis based on a case study of eight communities in the Taipei metropolitan area, Taiwan. Addressing both the temporal and spatial variations in rainfall, the GSBDS enhanced the broad application of RWHS evaluations. The results indicate that the scheme is feasible based on the optimal design when both water and energy-savings are evaluated. RWHSs were observed to be cost-effective and facilitated 21.6% domestic water-use savings, and 138.6 (kWh/year-family) energy-savings. Furthermore, the cost of per unit-energy-saving is lower than that from solar PV systems in 85% of the RWHS settings. Hence, RWHSs not only enable water-savings, but are also an alternative renewable energy-saving approach that can address the water-energy dilemma caused by rapid urbanization.

Keywords: economic feasibility; energy saving; geographic information system GIS; rainwater harvesting; water resources conservation

---

1. Introduction

Current centralized urban water supply systems depend largely on energy consumption in all processing phases, including purification, distribution, and sewage treatment [1,2]. However, only recently has attention been focused on exploring the connection between urban water supply and the associated energy consumption, known as the water-energy nexus [3–5]. In this context, the water-energy challenge is increasingly critical because of rapid urbanization [6–8]. Certain studies have also suggested that the synergistic impact of energy and water consumption portends more serious consequences if not addressed appropriately. In addition, future research should address challenges such as the environmental impact of urban water supply and energy production, and should develop models that jointly address energy and water conservation for the development of water and energy resilient cities [9–12]. Because rainwater is the most fundamental renewable resource, it can be harvested on-site and used for non-potable purposes (e.g., flushing toilets and gardening), without requiring complicated treatment and long-distance transportation. Therefore, the innovative conservation ability of rainwater harvesting systems (RWHSs) has the potential to ease the water-energy dilemma caused by rapid urbanization.

For many fast growing Asian cities with limited flat areas, such as Tokyo, Taipei, Seoul, and, Chongqing, water-energy concerns are further exacerbated, because the hilly areas surrounding the cities have often been developed into suburbs to accommodate fast growing urban populations [7,8,13]. Additional energy is required to pump water to these communities, which enlarges the ecological footprint [14,15]. To improve the water-energy efficiency of communities that live in hilly areas around large cities, attention should be focused on how to systematically and cost-effectively design RWHSs in a city scale. An integrated approach should be followed to shift focus from individual RWHSs to overall system performance. Research in this domain often uses historical rainfall data to simulate the hydraulic performance of individual RWHSs, predict economic feasibility, and seek optimal design [15–20]. The best known storage sizing for RWHSs is the water balance model. A review of the literature revealed that the water balance equation has been accepted and adopted by researchers to simulate the behavior of RWHSs, and to predict their performance [16,18,20–27]. RWHSs in dense urban areas appear to be economically advantageous [28–30]. To understand the energy-saving capacity of RWHSs, Retamal *et al.* [31]; Proença *et al.* [3]; Siddiqi and Anadon [32]; Abdallah and Rosenberg [33] have explored the energy saved due to RWHS system implementation, however, only single RWH sets are discussed. Because the frequency and amount of rainfall may also vary spatially, the primary disadvantage of a single-site approach lies in its improper treatment of the spatial aspect when large scale assessment is required [34]. To address the constraint attributed to both spatial and temporal variation of rainfall, some researchers used classified regions to present the spatial effect in the RWHS design [21,24,27]. To be more precise in the spatial analysis, Chiu and Liaw [34] incorporated water balance model into their geographic information system (GIS) and visualized the results, which resulted in a more comprehensive understanding of RWHSs. However, only the water-saving impact was considered and the water-energy nexus was not included in their analyses; consequently, the energy-saving effect and associated economic feasibility of RWHSs to enhance urban water-energy conservation remains unknown to the general public. Therefore, energy savings, spatial-based water balance model, and economic feasibility analysis should be carefully integrated. Only when communities and urban planners have received comprehensive and integrated information to improve their understating of RWHSs will they be able to make sound decisions when planning the water-energy conservation schemes of large cities.

This study established a GIS-simulation-based design system (GSBDS) that incorporated a historical rainfall data base, water balance model, spatial-based technologies, energy saving analysis, and economic feasibility analysis. This GSBDS case study was based on eight communities located in the hills around the Taipei metropolitan area in Northern Taiwan. The study's purpose was threefold: first, to perform a spatially-based simulation that addressed the temporal and spatial variation of rainfall together to enhance and visualize the performance of RWHSs; second, to obtain the optimal tank volumes of RWHSs and the associated total water and energy savings; and finally, to identify RWHS as an energy-saving approach for communities based in hilly areas by comparing them with another popular energy-saving approach, namely, solar photovoltaic (solar PV) system.

2. GSBDS Approach

2.1. System Description

The GSBDS consists of three parts, namely a data base, data processing, and data input and output, as illustrated in Figure 1. The data base included rainfall data, community data, and economic data, and data processing included water balance model, spatial interpolation, and economic feasibility analysis. The GSBDS application can be described as follows:

(1) The historical rainfall data of each rainfall station is first used for simulation based on water balance model to derive the water saving performance of RWHS in these stations. The results were then spatially interpolated to quantify and visualize the spatial distribution, and to predict annual water savings of the assigned RWHSs for communities based in hilly areas.

(2) Based on the water-energy coefficient derived from the community data and cost analysis of RWHSs, an economic feasibility analysis could be conducted to determine the cost-effectiveness of all RWHSs and thus identified optimal volumes of tanks.

(3) Using the optimized design of RWHSs, the total water-energy savings in the study area could be calculated, and the unit energy saving cost obtained to compare it with other energy saving approaches.

Water balance model and economic analysis were performed using VBA in SuperGIS 2.1 and SQL to access the data base [34]. The details of the water balance model, spatial interpolation, water-energy relation, and economic feasibility analysis are described below.

2.2. Spatial Interpolation

Various methods for spatial interpolation, including Kriging, Tend, Spline, Thissen Polygon, Multivariate Regression, and Inverse Distance Weighted (IDW) have been developed to estimate the value of properties at sites without observational data [35]. However, none of these has been accepted as the standard method. In Taiwan, the IDW ($p = 2$) method was reported effective to interpolate rainfall data in Northern Taiwan, and was therefore adopted for this study [36]. The IDW was calculated using the GIS software, SuperGIS 2.1 [37].

2.3. Water Balance Model

For most RWHSs, the efficiency of rainwater supply depends on four major parameters, namely, rainfall depth, rooftop area, tank volume, and water demand. Our GSBDS incorporated the Yield-Before-Spillage (YBS) release rule with time step of daily [38], which determines the demand withdrawn before spillage.

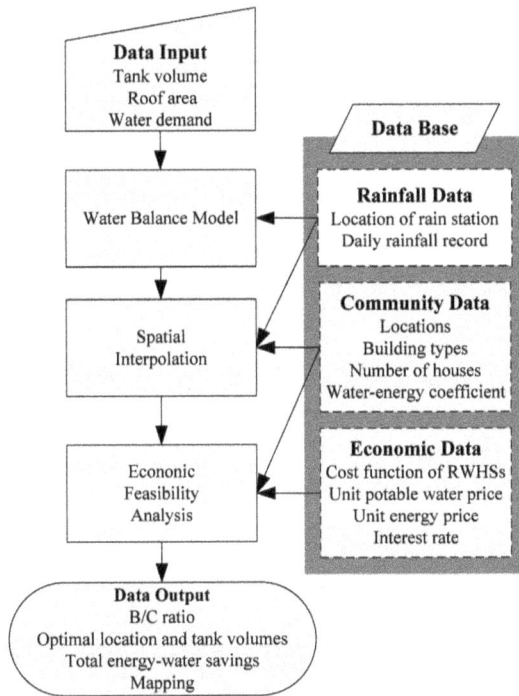

Figure 1. Conceptual framework of GSBDS.

The operational algorithm of YBS can be mathematically described as follows:

$$Y_t = Min(D_t, S_{t-1}) \tag{1}$$

$$S_t = Min(S_{t-1} + Q_t - Y_t, V) \tag{2}$$

$$Q_t = C \cdot I_t \cdot A \tag{3}$$

where Y_t is the rainwater yield (m^3) during the tth period; D_t is water demand (m^3) at time t; S_{t-1} is storage volume (m^3) of rainwater in the tank at the t-1th period; V denotes tank volume (m^3); Q_t denotes the rooftop rainwater runoff (m^3); I_t is rainfall depth (m) at time t; A is catchment area (m^2); and C is the runoff coefficient (runoff/rainfall volume). (setting 0.82 as recommended by Liaw and Tsai [17]).

Based on the design factors (*i.e.*, A, D_t, V) and the rainfall data of a rainfall station, the hydraulic performance (annual potable water savings WS (m^3/year) of the RWHS of a rainfall station could be achieved and can be expressed as follows:

$$WS = \frac{\sum\limits_{t-1}^{n} Y_t}{n} 365 \tag{4}$$

where n denotes the total number of daily rainfall data. To estimate the WS of various communities (WSi of ith community), a spatial interpolation method based on the WS of each rainfall station was subsequently adopted.

51

2.4. Water-Energy Nexus

Before exploring the economic feasibility of RWHSs for water-energy savings, the relationship between potable water consumption and energy consumption, namely water-energy coefficient β (kWh/m^3), had to be assessed. Typically, potable water from purifying plants is first pumped by urban pump stations supplying water to both flat and hilly areas, and then pumped by a series of hilly area pump stations to communities based in the area. Cheng [1] investigated the water-energy coefficients in downtown areas of Taipei City, and reported an average of 0.22 kWh/m^3 for water purifying, 0.17 kWh/m^3 for urban pumping, and 0.41 kWh/m^3 for sewage treatment. However, because of the difficulties involved in collecting data, water-energy coefficients for pumping in hilly areas were unknown and required further investigation.

The amount of energy required to pump water depends on a number of factors, including lift, water flow, pump efficiency, transmission efficiency, and friction-loss of pipes, and can be simplified by measuring the power of pumps and the water flow [39]. Each participating community's water-energy coefficient for pumping in hilly areas can be mathematically defined as follows:

$$\beta_{Hilly,i} = \sum_{j} \beta_{Hilly,i,j} = \sum_{j} \frac{R_{p,i,j} \cdot 0.746 \cdot (1/\eta)}{Q_{p,i,j} \cdot (1/24)} \tag{5}$$

where $\beta_{Hilly,i}$ is the water-energy coefficient of ith hilly community; $\beta_{Hilly,i,j}$ denotes the water-energy coefficient for the jth hilly pump station in i^{th} community; $R_{p,i,j}$ is pump power (hp) in j^{th} hilly pump stations in i^{th} hilly community; η is mechanical efficiency (%), and $Q_{p,i,j}$ is the water flow (m^3/d).

The cost of purifying water and urban pumping are typically included in water fees, and are therefore regarded as external effects in the economic analysis. In addition, applying RWHSs to flushing toilets and gardening cannot reduce the amount of water used for sewage treatment. Therefore, the energy cost for sewage treatment was also excluded from the economic analysis presented in the next section. However, when the total energy savings and the associated CO_2 emissions are required to be considered, the water-energy coefficient for water purifying ($\beta_{Purifing}$) and urban pumping of associated i^{th} community ($\beta_{Urban,i}$) have to be included. In this case, the total water-energy coefficient of ith community ($\beta_{Total,i}$) can be expressed as follows:

$$\beta_{Total,i} = \beta_{Purifing} + \beta_{Urban,i} + \beta_{Hilly,i} \tag{6}$$

The operational energy used in RWHSs may vary according to the types of RWHS design. Vieira *et al.* (2014) [40] reviewed literatures and reported that the energy consumption per unit rainwater supply may be far higher than that in centralized urban water systems when direct pumps are adopted to supply stored rainwater to end users. This is because the energy used for pump start-ups and standby mode is often underestimated. However, the energy efficiency of pumps can be enhanced when RWHSs are based on gravity rainwater supply by using rainwater header tanks [40]. Rainwater from roofs is first collected in the storage tanks on the ground, and then pumped to the header tanks that usually are placed on the roofs or on the higher position in the buildings. The header tanks supply water via gravity and are commonly installed with dual-inlet float valves that supply potable water when rainwater storage tanks are empty. In Taiwan, the header tank's design is common and widely adopted to store potable water. This study also adopted RWHSs designed with rainwater header tanks, and thereby assumed the operational energy consumption of RWHSs to be negligible in the case study, because the energy used for pumping stored rainwater to header rainwater tanks can be compensated by reducing pumping water to the potable header tanks.

2.5. Economic Feasibility Analysis and the Optimal Design

The economic feasibility of this RWHS was determined by its benefit-cost (B/C) ratio, where a B/C ratio greater than 1 was considered cost-effective. To simplify the calculation, the cost and benefit of RWHS were both expressed in unit form, as well as in New Taiwan Dollars (US$1 = NT$32.5). The cost of unit water savings ($UWSC_{i,v}$, NT$/m³, for i^{th} community using V tank) and the cost of unit energy saving ($UESC_{i,v}$, NT$/kWh, for i^{th} community using V tank volume) could be derived from the results of GSBDS data processing (Figure 1), using water balance model, spatial interpolation, and Equation (5). The $UWSC_{i,v}$ and $UESC_{i,v}$ can be written as follows:

$$UWSC_{i,v} = \frac{C_v}{WS_{i,v}} \tag{7}$$

$$UESC_{i,v} = \frac{C_v}{\beta_{Hilly,i} \cdot WS_{i,v}} \tag{8}$$

where C_v is the annualized cost of RWHS, using V tank volume; and $WS_{i,v}$ denotes annual potable water saved in i^{th} community, using V tank volume.

A larger tank volume implies more water saved and a greater contribution to energy savings, but is less cost-effective. Therefore, we assumed that the largest tank volume that remained cost-effective to be the optimal design for the GSBDS.

Customarily, water conservation and energy conservation are rarely planned together. For this reason, the B/C ratios of ith community with V tank volume were examined in terms of three scenarios: Scenario 1 considered only the water-saving benefit "$(B/C)_{W,i,v}$"; Scenario 2 considered only the energy-saving benefit "$(B/C)_{E,i,v}$"; and Scenario 3 considered both water and energy savings "$(B/C)_{W+E,i,v}$":

Scenario 1:

$$\left(\frac{B}{C}\right)_{W,i,v} = \frac{UWF}{UWSC_{i,v}} \tag{9}$$

Scenario 2:

$$\left(\frac{B}{C}\right)_{E,i,v} = \frac{UEF}{UESC_{i,v}} \tag{10}$$

Scenario 3:

$$\left(\frac{B}{C}\right)_{W+E,i,v} = \frac{UWF + \beta_{Hiily,i} \cdot UEF}{UWSC_{i,v}} \tag{11}$$

where UWF denotes the unit potable water fee ($/m³), and UEF is the unit electricity fee ($/kWh). The UWF and UEF were considered as benefits of unit savings in this analysis.

Based on this optimal design, the contribution of applying RWHS to a water-energy conservation scheme could be described using the total potable water savings per year TWS (m³/year), and the total energy savings per year TES (kWh/year):

$$TWS = \sum WS_i \cdot N_i \tag{12}$$

$$TES = \sum WS_i \cdot \beta_{Total,i} \cdot N_i \tag{13}$$

where N_i is the number of buildings of selected building type in ith community.

3. Background and Data Sources

Taipei is the largest metropolitan area in Taiwan and includes both Taipei City and New Taipei City which is located in the Taipei Basin, and is well known for its large population of 6.36 million (in year 2014). Because Taipei's development is constrained by a limited plain area, government and private enterprises have launched several large community projects in the surrounding hills since the early 1980s. As a case study, rooftop RWHSs were assumed in these communities and applied in the GSBDS. The rainfall data, RWHSs cost, and community data can be described as follows.

3.1. Rainfall Data

The rainfall data base included the daily rainfall data from 31 automatic rain data collection stations (15 inside and 16 around Taipei City, collected from 1994 to 2013). Mitchell *et al.* [41] and Campisano *et al.* [42] have suggested that the favorable rainfall data set consistency is at minimum 25 years for making a reliable simulation. However, in this case study, out of 31 rainfall stations, some of them do not provide complete and reliable rainfall data before 1994. Based on the suggestion from previous study and the reality of available data set, this study adopt 20 years of rainfall data for simulation in a large scale. Among them, the average annual rainfall was 3127 mm/year, the lowest was 1321 mm/year, and the highest was 6124 mm/year. Despite an annual rainfall difference of nearly 4.6 times, the distance between these two stations is only 17.2 km, revealing how substantially the spatial factor affects the performance of RWHSs, and highlights the importance of adopting a spatial approach. Figure 2 indicates the locations of the rainfall stations and the spatial variances of average annual rainfall where the northeastern area is higher and northwestern area is lower.

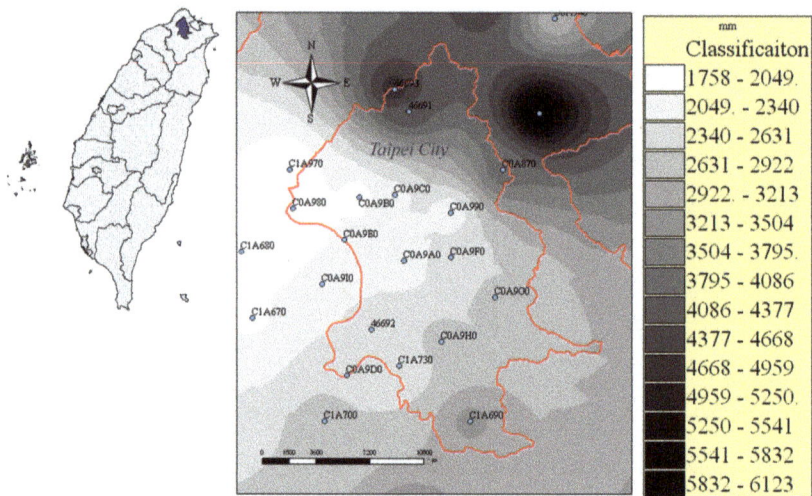

Figure 2. Average annual rainfall and the locations of rainfall stations.

3.2. Water Quality Issue and the Cost of RWHSs

The typical rooftop RWHS$_S$ consists of four basic sub-systems, namely a catchment system, pipeline system, treatment system, and tank system. The level of treatment to achieve acceptable water quality might substantially influence the cost of RWHSs. In Taiwan, the Water Resources Agency (WRA) established a recommended contaminant level for rainwater harvested from rooftops to flush toilets [43]. Because biological oxygen demand (BOD) and free chloride surplus are not regulated, appropriately installed RWHSs with simple filters typically meet the rainwater reuse standard.

Chiu *et al.* [38,44] surveyed market prices to develop a cost function for RWHSs in Northern Taiwan using the judgment sampling technique. We adopted the most commonly used cylindrical stainless steel tank, with a 25-year lifespan; 5% of the total cost for maintenance; and 3% discount rate which is recommended by the Taiwanese government for evaluating public projects. The annualized cost function $C(V)$ of RWHSs of this type of tank can be expressed as follows:

$$C(V) = 944.65 + 34.10 \cdot V^2, V \geq 1, R^2 = 0.997 \tag{14}$$

where V (m^3) is tank volume.

3.3. Community Data

Among all the communities based in the hilly areas of the Taipei metropolitan area, hilly area pump stations in eight communities are maintained by the main public utility operator, the Taipei Water Department (TWD). The other pump stations are maintained by private construction companies or associated community committees. In this case study, pump station data from the eight hill communities maintained by the main public utility were analyzed and adapted to the GSBDS.

Table 1 lists details of the eight communities that formed part of this study. A sequence number was assigned according to the respective average annual rainfall obtained by interpolating the data from rainfall stations.

Table 1. Details of hilly communities with pumping stations managed by TWD.

i	Community	N_i	Rainfall (mm)	Longitude	Latitude	$\beta_{Urban,i}$	$\beta_{Hilly,Lj}$			
							$j = 1$	$j = 2$	$j = 3$	$j = 4$
1	Yie She	597	2744	25°00′27.21″	121°34′01.31″	0.22	0.338	0.573	–	–
2	Rose China	720	2867	24°56′46.06″	121°34′43.12″	0.17	0.331	0.166	0.197	–
3	Taipei small town	1418	2911	24°56′44.40″	121°29′59.90″	0.17	0.191	0.406	1.043	0.796
4	Wan Fan	225	2941	25°00′04.85″	121°34′08.45″	0.17	0.358	0.177	–	–
5	Ju Chun Li	239	3031	25°01′50.21″	121°39′57.94″	0.22	0.249	0.904	0.266	
6	Chi Nan	152	3059	24°59′20.88″	121°35′07.47″	0.22	0.628	0.628	0.419	0.266
7	Tan shi Shan	116	3108	24°56′53.79″	121°31′59.90″	0.22	0.744	–	–	–
8	Mu Ja 2 phase	585	3255	24°56′20.88″	121°35′07.47″	0.22	0.331	0.124	–	–

N_i is the number of buildings of selected building type in i^{th} community, which also refers to the total number of RWH systems. Because various types of buildings exist in these communities, the most common types of building, a building with a 100-m^2 rooftop area supplying water for a family's toilet flushing, and a 20-m^2 garden area, were selected for this study to further simplify the calculations.

Another critical design consideration was water demand. Based on the government's 250 L per capita per day suggestion (Lpcd) for indoor water consumption, of which 24% is used for toilet flushing in a typical four-member household [45]. The gardening water demand was set to 3.0 mm/d based on the recommendation of the Technical Guideline for Soil and Water Conservation [46].

4. Results and Discussion

4.1. Results of Simulation

The results of the spatially interpolated WS and the locations of the hill communities are presented in Figure 3. The spatial variation of WS agreed with the spatial distribution of average annual rainfall as demonstrated in Figure 2; however, further analysis is required to understand the performance of RWHSs on a large scale. Several previous studies used the generalized method that simply applies a constant usage rate to the average annual or monthly rainfall data without addressing the temporal variation of rainfall [47,48]. Figure 4 clearly shows the inconsistent of average annual rainfall and WS_i of each community using most commercially available tank volumes, *i.e.*, 1, 3 and 5 m^3. This discrepancy occurred because of the difference in rainfall patterns. For example, despite similar average annual rainfall, RWHSs in areas with an uneven temporal distribution of rainfall tended to be less efficient than those with even distribution, because excess rainwater in a single rainfall event eventually overflows without being used. Therefore, the GSBDS used both water balance model and spatial technology to yield more precise information for the economic feasibility analysis.

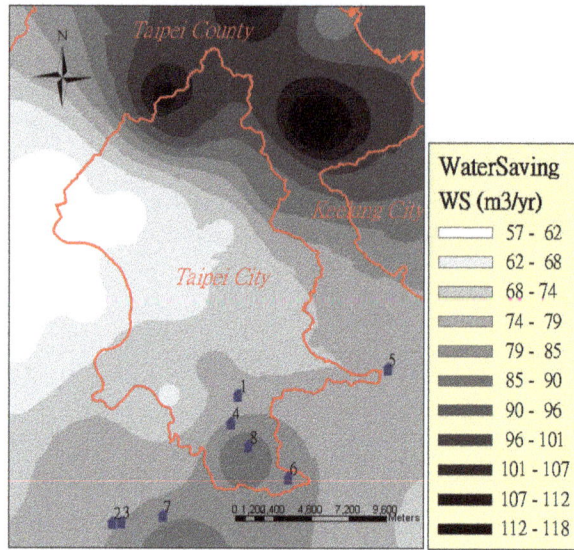

Figure 3. The locations of the communities, and the spatial variation of *WS* (m³/year) (assuming $A = 90$ m², $V = 3$ m³) per household and rainwater used for toilet flushing and gardening.

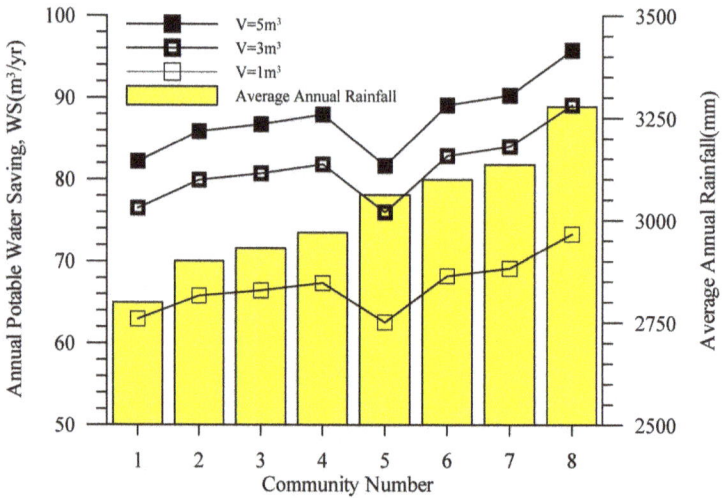

Figure 4. The average annual rainfall and *WS* of each community.

4.2. Economic Feasibility Analysis and Optimal Design

Figure 5 depicts and compares the results of the three scenarios used in the economic feasibility analysis of each community. This clearly indicates that all RWHSs of Scenarios 1 and 2 that considered only the benefits of water-saving, or only the benefits of energy-saving were unfeasible. This poor economic performance is partially because the Taiwanese government currently subsidizes water and electricity. In Scenario 3, which considered both water and electricity benefits combined, all RWHSs with smaller tank volumes were cost effective, and are therefore a feasible water-energy conservation scheme for hill communities. The findings of this study therefore demonstrate that considering energy and water use together provides valuable insights that do not arise from discrete analyses of only water or only energy.

Additionally, Scenario 3 in Figure 5 shows that the optimal tank volumes for each community could be obtained and are listed in Table 2. The approach demonstrated in Figure 5 is a simple and straightforward method to support users to determine the optimal RWHSs design on a large scale. Table 2 also includes the related WS_i and energy savings factors required to calculate the total water-energy savings.

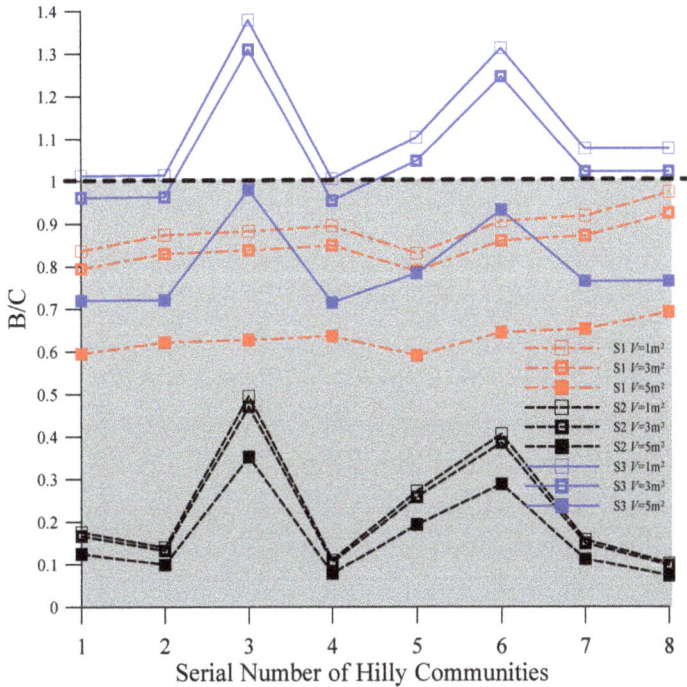

Figure 5. Results of economic feasibility analysis (UWF = 13NT\$/m^3, UEF = 3NT\$/kWh).

Table 2. The optimal tank volumes and related water-energy saving details.

i	Optimal Volume (m^3)	Water Savings (m^3/year)	Energy Savings (kWh/year)	Unit Energy Saving Cost (NT\$/kWh)
1	1	62.9	85.0	17.1 *
2	1	65.8	71.3	21.5 *
3	3	80.7	228.1	6.4 *
4	1	67.3	62.3	27.2 *
5	3	76.0	141.3	11.7 *
6	3	82.9	197.4	7.8 *
7	3	84.0	99.5	20.0 *
8	3	89.1	79.7	30.8 *

* denotes the $UESC$ is lower than the one of solar PV systems.

4.3. Contribution to Water-Energy Conservation

Based on the optimal design, the total water savings should reach 307,173 m^3/year (*i.e.*, 75.8 m^3 per family per year). The total energy savings amounts to 561,438 (kWh/year) (*i.e.*, an average of 138.6

Water **2015**, *7*, 6285–6300

(kWh) per family per year). Households that use RWHSs could expect water savings equivalent to an average of 21.3% of domestic water use, or an energy saving of 21.3% for water pump costs in hilly areas.

4.4. Comparison of RWHS with Solar Energy

Growing awareness of the energy crisis and the effects of global warming are making solar PV systems increasingly popular in Taiwan, but the energy saving potential of RWHSs remains unknown to the public. Therefore, commercially available solar PV systems were selected for comparison. In Taipei, the *UESC* of installing PV systems has been reported to range from 28 to 30 NT\$/kWh [49]. The *UESC* of RWHSs was calculated and is listed in Table 2. It ranged from 6.4 to 30.8 NT\$/kWh. The *UESC* of RWHSs in 85.5% of households, that is, seven communities out of eight, is lower than that of solar PV. Therefore, before the cost of PV panels decreases substantially, harvesting rainwater should be regarded as a feasible option to address the emerging water-energy demand of communities based in hilly areas.

The GSBDS is a useful tool to evaluate and design RWHSs for water-energy conservation, and this case study demonstrates that RWHSs are feasible when both water and energy savings are considered. However, the feasibility and benefits of applying RWHS may be underestimated if only direct monetary benefit is considered. Indirect benefits, such as CO_2 emission reduction, storm runoff mitigation, reducing water stress during peak hours, and decreased demand on current water and energy facilities, should not be overlooked when planning RWHS for urban development. This case study's data may not be applicable to all communities of the Taipei metropolitan area. Therefore, more community, building, and RWHS types should be considered in future studies before RWHSs can be fully implemented as an essential component of a water-energy conservation scheme for large cities.

5. Conclusions

Water-energy challenges can constrain ongoing development in many rapid growing cities. Therefore, a system to conserve rooftop rainwater in the hill communities surrounding the Taipei metropolitan area was systematically designed and evaluated in this study. We established a GSBDS to identify the hydraulic performance of RWHSs, quantify the economic feasibility of RWHSs for optimal design, and understand RWHSs water-energy saving potential.

Based on the spatial technology and water balance model of the GSBDS, the accuracy of evaluation was enhanced, and the hydraulic performance of RWHSs was quantified for economic analysis. Traditionally, water conservation efforts seldom incorporate energy conservation, however, the economic feasibility findings of this study strongly suggest that the economic feasibility is substantially enhanced when both water and energy savings are considered together. The results also revealed that, based on the optimal RWHS design derived from the economic feasibility analysis, the contribution of RWHSs to water-energy conservation can reach 307,173 m^3/year (*i.e.*, 75.8 m^3 per family per year) and 561,438 kWh/year (*i.e.*, an average of 138.6 kWh per family per year); which is an average of 21.3% savings in domestic water use, or a 21.3% energy savings in water pumping in hilly areas. Additionally, RWHSs are more cost-effective than solar PV systems for 85.5% of households in hill communities.

RWHSs are therefore not only a practical water-saving strategy, but also a feasible energy-saving approach. One of the advantages of a GSBDS is its ability to provide a systematic and comprehensive method to explore the energy and water savings of RWHSs on a city scale. Therefore, GSBDS is a useful tool to evaluate and design RWHSs to ease the growing water-energy shortage dilemma caused by rapid urbanization.

Acknowledgments: The authors wish to acknowledge the essential support and from Taipei Water Department and Jin Win University of Science and Technology.

Author Contributions: All authors were involved in designing and discussing the study. Yie-Ru Chiu executed the model and coordinated the group. Yao-Lung Tsai and Yun-Chih Chiang collected required data and finalized the manuscript. All authors have read and approved the final manuscript.

Conflicts of Interest: The authors declare no conflict of interest.

References

1. Cheng, C.L. Study of the inter-relation between water use and energy conservation for a building. *Energy Build.* **2002**, *34*, 261–266. [CrossRef]
2. Jothiprakash, V.; Sathe, M.V. Evaluation of rainwater harvesting methods and structures using analytical hierarchy process for a large scale industrial area. *J. Water Resour. Prot.* **2009**, *1*, 427–438. [CrossRef]
3. Proenca, L.C.; Ghisi, E.; da Fonseca Tavares, D.; Coelho, G.M. Potential for electricity savings by reducing potable water consumption in a city scale. *Resour. Conserv. Recycl.* **2011**, *55*, 960–965. [CrossRef]
4. Scott, C.A.; Pierce, S.A.; Pasqualetti, M.J.; Jones, A.L.; Montz, B.E.; Hoover, J.H. Policy and institutional dimension of the water-energy nexus. *Energy Policy* **2011**, *39*, 6622–6630. [CrossRef]
5. Plappally, A.K.; Lienhard, V.J.J. Energy requirement for water production, treatment, end use, reclamation, and disposal. *Renew. Sustain. Energy Rev.* **2012**, *16*, 4818–4848. [CrossRef]
6. Varis, O.; Somlyody, L. Global urbanization and urban water: Can sustainability be afforded? *Water Sci. Technol.* **1997**, *35*, 21–32. [CrossRef]
7. Douglass, M. *East Asian Urbanization: Patterns, Problems, and Prospect*; Asia/Pacific Research Center, Stanford University: Stanford, CA, USA, 1998.
8. Asian Development Bank (ADB). *Thinking about Water Differently, Managing the Water-Energy-Food Nexus*; Asian Development Bank: Mandaluyong, Philippines, 2013.
9. Hightower, M.; Cameron, C.; Pate, R.; Einfeld, W. Emerging energy demands on water resources. *Water Resour. Impact* **2007**, *9*, 8–11.
10. Phoenix, L.E. Introduction: Energy and water. *Water Resour. Impact* **2007**, *9*, 1–3.
11. Talebpour, M.R.; Sahin, O.; Siems, R.; Stewart, R.A. Water and energy nexus of residential rain water tanks at an end use level: Case of Australia. *Energy Build.* **2014**, *80*, 195–207. [CrossRef]
12. Farreny, R.; Gabarrell, X.; Rieradevall, J. Cost-efficiency of rainwater harvesting strategies in dense Mediterranean neighborhoods. *Resour. Conserv. Recycl.* **2011**, *55*, 686–694. [CrossRef]
13. World Bank. Delivering water to Mexico City. *Urban Age Mag.* **1999**, *6*, 16–17.
14. Rees, W.E. Ecological footprints and appropriated carrying capacity: What urban economics leaves out. *Environ. Urban.* **1992**, *4*, 121–130. [CrossRef]
15. Beal, C.; Bertone, E.; Stewart, R.A. Evaluating the energy and carbon reductions resulting from resource-efficient household stock. *Energy Build.* **2012**, *55*, 422–432. [CrossRef]
16. Fewkes, A. The use of rainwater for WC flushing: The field testing of a collection system. *Build. Environ.* **1999**, *34*, 765–772. [CrossRef]
17. Liaw, C.H.; Tsai, Y.L. Optimum storage volume of rooftop rainwater harvesting systems for domestic use. *J. Am. Water Resour. Assoc.* **2004**, *40*, 901–912. [CrossRef]
18. Villarreal, E.L.; Dixon, A. Analysis of a rainwater collection system for domestic water supply in Ringdansen, Norrkoping, Sweden. *Build. Environ.* **2005**, *40*, 1174–1184. [CrossRef]
19. Ghisi, E. Potential for potable water savings by using rainwater in the residential sector of Brazil. *Build. Environ.* **2006**, *41*, 1544–1550. [CrossRef]
20. Chisi, E.; Ferreira, D.F. Potential for potable water saving by using rainwater and greywater in a multi-storey residential building in southern Brazil. *Build. Environ.* **2007**, *42*, 2512–2522.
21. Liaw, C.H.; Chiang, Y.C. Dimensionless analysis for designing domestic rainwater harvesting systems at the regional level in northern Taiwan. *Water* **2014**, *6*, 3913–3933. [CrossRef]
22. Imteaz, M.A.; Ahsan, A.; Shanableh, A. Reliability analysis of rainwater tanks using daily water balance model: Variations within a large city. *Resour. Conserv. Recycl.* **2013**, *77*, 37–43. [CrossRef]
23. Imteaz, M.A.; Matos, C.; Shanableh, A. Impacts of climatic variability on rainwater tank outcomes for an inland city, Canberra. *Int. J. Hydrol. Sci. Technol.* **2014**, *4*, 177–191. [CrossRef]
24. Imteaz, M.A.; Rahman, A.; Ahsan, A. Reliability analysis of rainwater tanks: A comparison between South-East and Central Melbourne. *Resour. Conserv. Recycl.* **2012**, *66*, 1–7. [CrossRef]
25. Fewkes, A.; Butler, D. Simulating the performance of rainwater collection and reuse systems using behavioural models. *Build. Serv. Eng. Res. Technol.* **2000**, *21*, 99–106. [CrossRef]

26. Campisano, A.; Modica, C. Appropriate resolution timescale to evaluate water saving and retention potential of rainwater harvesting for toilet flushing in single houses. *J. Hydroinformatics* **2015**, *17*, 331–346. [CrossRef]

27. Campisano, A.; Gnecco, I.; Modica, C.; Palla, A. Designing domestic rainwater harvesting systems under different climatic regimes in Italy. *Water Sci. Technol.* **2013**, *67*, 2511–2518. [CrossRef] [PubMed]

28. Matos, C.; Santos, C.; Pereira, S.; Bentes, I.; Imteaz, M. Rainwater storage tank sizing: Case study of a commercial building. *Int. J. Sustain. Built Environ.* **2013**, *2*, 109–118. [CrossRef]

29. Hajani, E.; Rahman, A. Reliability and cost analysis of a rainwater harvesting system in peri-urban regions of Greater Sydney, Australia. *Water* **2014**, *6*, 945–960. [CrossRef]

30. Matos, C.; Santos, C.; Bentes, I.; Imteaz, M.A.; Pereira, S. Economic analysis of a rainwater harvesting system in a commercial building. *Water Resour. Manag.* **2015**, *29*, 3971–3986. [CrossRef]

31. Retamal, M.; Glassmire, J.; Abeysuriya, K.; Turner, A.; White, S. *The Water-Energy Nexus: Investigation into the Energy Implications of Household Rainwater Systems*; Institute for Sustainable Futures, University of Technology Sydney: Sydney, Australia, 2009.

32. Siddiqi, A.; Anadon, L.D. The water-energy nexus in Middle East and North Africa. *Energy Policy* **2011**, *39*, 4529–4540. [CrossRef]

33. Abdallah, A.; Rosenberg, D. Heterogeneous residential water and energy linkages and implications for conservation and management. *J. Water Resour. Plan. Manag.* **2014**, *140*, 288–297. [CrossRef]

34. Chiu, Y.R.; Liaw, C.H. Designing rainwater harvesting systems for large-scale potable water saving using spatial information system. *Lect. Note Comput. Sci.* **2008**, *5236*, 653–663.

35. Wei, H.; Li, J.L.; Liang, T.G. Study on the estimation of precipitation resources for rainwater harvesting agriculture in semi-arid land of China. *Agric. Water Manag.* **2005**, *71*, 33–45. [CrossRef]

36. Hsieh, H.H.; Chengm, S.J.; Liou, Y.J.; Chou, S.C.; Siao, B.R. Characterization of spatially distributed summer daily rainfall. *J. Chin. Agric. Eng.* **2006**, *52*, 47–55. (In Chinese)

37. SuperGeo Technologies Inc. *User's Menu of SuperGIS*; SuperGeo Technologies Inc.: Taipei, Taiwan, 2004.

38. Chiu, Y.R.; Liaw, C.H.; Hu, C.Y.; Tsai, Y.L.; Chang, H.H. Applying GIS-based rainwater harvesting design system in the water-energy conservation scheme for large cities. In Proceedings of the IEEE 13th International Conference on Computer Supported Cooperative Work in Design (CSCWD 2009), Santiago, Chile, 22–24 April 2009; pp. 722–727.

39. McGhee, T.J. *Water Supply and Sewerage*; McGraw-Hill: Singapore, 1991.

40. Vieira, A.S.; Beal, C.D.; Ghisi, E.; Stewart, R.A. Energy intensity of rainwater harvesting systems: A review. *Renew. Sustain. Energy Rev.* **2014**, *34*, 225–242. [CrossRef]

41. Mitchell, V.G.; McCarthy, D.T.; Deletic, A.; Fletcher, T.D. Urban stormwater harvesting-sensitivity of a storage behaviour model. *Environ. Model. Softw.* **2008**, *23*, 782–793. [CrossRef]

42. Campisano, A.; Modica, C. Regional scale analysis for the design of storage tanks for domestic rainwater harvesting systems. *Water Sci. Technol.* **2012**, *66*, 1–8. [CrossRef] [PubMed]

43. Water Resources Agency, Ministry of Economic Affairs. Available online: Http://www.wra.gov.tw/ct.asp?xItem=34751&ctNode=6259&comefrom=lp (accessed on 12 April 2014).

44. Chiu, Y.R.; Liaw, C.H.; Chen, L.C. Optimizing rainwater harvesting systems as an innovative approach to saving energy in hilly communities. *Renew. Energy* **2009**, *34*, 492–498. [CrossRef]

45. Architecture and Building Research Institute (ABRI). *Evaluation Manual for Green Building in Taiwan*; Ministry of the Interior: Taipei, Taiwan, 2015. (In Chinese)

46. Soil and Water Conservation Bureau (SWCB). *Technical Guidelines of Water and Soil Conservation*; Council of Agriculture: Taipei, Taiwan, 2014. (In Chinese)

47. Wung, T.C.; Lin, S.H.; Hung, S.M. Rainwater reuse supply and demand response in urban elementary school of different districts in Taipei. *Resour. Conserv. Recycl.* **2006**, *46*, 149–167. [CrossRef]

48. RELMA in ICRAF & UNEP. Rainwater Harvesting Potential in Africa: A GIS Overview Volume One. Available online: Http://www.unep.org/pdf/RWH_in_Africa-final.png (accessed on 15 July 2014).

49. Architecture and Building Research Institute (ABRI). *Final Report of Study on Solar Radiate Criterion for PV Design and Related Testing Standard*; Ministry of the Interior: Beijing, China, 2005. (In Chinese)

water

MDPI

Article

Identification of Decisive Factors Determining the Continued Use of Rainwater Harvesting Systems for Agriculture Irrigation in Beijing

Xiao Liang [1,*] and Meine Pieter van Dijk [2]

[1] School of Economics, Shenzhen University, Guangdong 518060, China
[2] International Institute of Social Studies, The Hague 2518AX, The Netherlands; mpvandijk@iss.nl
*   Correspondance: x.liang@yahoo.com or liangx@szu.edu.cn; Tel.: +86-755-2673-3256; Fax: +86-755-2653-5344

Academic Editor: Ataur Rahman
Received: 14 October 2015; Accepted: 16 December 2015; Published: 25 December 2015

Abstract: The success or failure of operating a rainwater harvesting system (RWH) depends on both technological and non-technological factors. The importance of non-technological factors in attaining sustainable RWH operation is rarely emphasized. This study aims to assess the contribution of non-technological factors through determining decisive factors involved in the use of RWHs for agriculture irrigation in Beijing. The RWHs for agriculture irrigation in Beijing are not operating as well as expected. If the decisive factors are identified to be non-technological, the significance of non-technological factors will be highlighted. Firstly, 10 impact factors comprising non-technological and technological factors are selected according to both a literature review and interviews with RWH managers. Following this, through an artificial data mining method, rough set analysis, the decisive factors are identified. Results show that two non-technological factors, "doubts about rainwater quality" and "the availability of groundwater" determine whether these systems will continue or cease RWH operation in Beijing. It is, thus, considered necessary to improve public confidence in and motivation on using rainwater for agriculture irrigation, as this is the main obstacle in the sustainable and successful operation of RWHs. Through a case study of RWHs in Beijing, the study verifies the importance of acknowledging non-technological factors to achieve sustainable water management and considers that such factors should receive more attention by decision makers and researchers.

Keywords: rainwater harvesting; sustainable water management; decisive factors; rough set analysis

---

1. Introduction

Factors considered to be important in achieving sustainable water management vary amongst the disciplines of engineering, economics, and social sciences. From the respective of technical improvements, the impact factors on rainwater harvesting systems (RWH) include optimal tank size, technical design, and so on [1–5]. Technical improvements could effectively facilitate the operation of a new system, but they do not determine its successful operation. Whether the operation of a system is a success or failure depends on both technological and non-technological factors. Non-technological factors can become considerable obstacles in the adequate functioning of RWHs. Discussions in literature pertaining to non-technological factors in RWHs management contain economic analysis, public perception, and so on [6–10]. However, although these studies have analyzed and explained how factors influence RWHs in a scientific manner, it is rare to emphasize the importance of non-technological factors in attaining sustainable RWHs operation. It is also considered that in some management settings, non-technological factors are more critical than technological ones. Thus, this study aims to assess the contribution of non-technological factors to the efficient running of RWHs.

The assessment is implemented through determining decisive factors involved in the use of RWHs in a rural area of Beijing. There is scarcity of water in Beijing because of the large population, continual drought, and depletion of groundwater stocks, and water used for agricultural irrigation accounts for around 60% of the total water use in Beijing. To ameliorate problems associated with this lack of available water, hundreds of RWHs have been constructed for use in agricultural irrigation since the year 2006, which were largely promoted and subsidized by the government. However, most of these systems are actually not operating as well as expected [7] and it is thus very important to determine decisive factors influencing this in Beijing. In the literature, there are studies in discussing determinants of RWHs [11–13], which mostly pursue to find the factors influencing the adoption of RWHs. While, this study emphasizes to prove that non-technological factors are as substantial as technological factors in the successful management of RWHs via the identification of decisive factors.

An artificial data mining method, rough set analysis, was carried out. The sample in this study is small, and the selected impact factors are obtained using partly qualitative and partly quantitative data. Rough set analysis is a mathematical method used to synthesize an approximation of concepts from data that allows information classification [14], even though sample size is small. An extensive theoretical description is provided by Pawlak (1982) and Slowinski (1991) [15,16]. The method has rarely been applied to water resource studies [14,17], although it has been applied in other scientific fields such as medicine, economics, software engineering, and urban studies [18–21].

Firstly, non-technological and technological factors were selected according to both a literature review and interviews with RWH managers. Following this, through the method of rough set analysis, impact factors that were decisive in determining the operational success or failure of RWH were identified. If these factors were found to be non-technological, their importance in successful operations were highlighted.

2. Methodology

2.1. Data

An extensive field study was conducted in a collaborative project between the Chinese Academy of Sciences and the Beijing Agro-Technical Extension Center (BATEC), which focused on RWH practices in the rural areas of Beijing. Some plants are supervised and subsidized by BATEC, which is a professional institute involved with agricultural technology, but others are supervised by other institutes, such as the Beijing Water Saving Office. There is limited access to the plants supervised by the other institutions, and thus the fieldwork was focused only on RWHs supervised by BATEC. The selection criteria determined that each plant used in the study had to have been constructed at least 2 years prior to the study, and that sufficient data on the plant was available. Therefore, only 10 RWHs were selected for study. Although the sample is small, complete data are available for each plant. Locations of the 10 plants are shown in Figure 1, which are distributed within six of Beijing's districts.

The collected rainwater in these RWHs is generally reused for greenhouse irrigation. Figure 2 shows the structure of a typical RWH system. Rainwater goes through the plastic film which is covering the greenhouse down to the ditch in front of the greenhouse, and then moves to the sediment tank for filtering through a big underground pipe. After the solids are deposited, cleaned water enters into the storage tank. When irrigation is required, water is pumped from the storage tank to the greenhouse.

Figure 1. Locations of RWHs studied within Beijing.

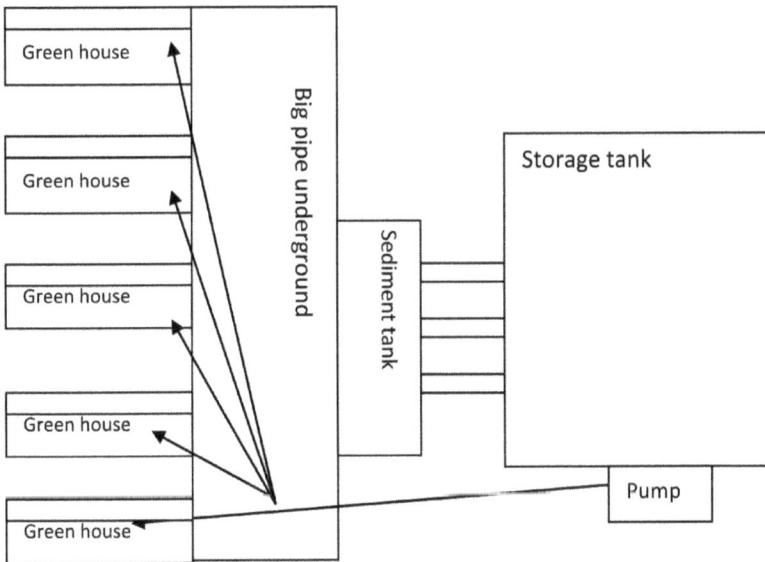

Figure 2. A RWH system [7].

2.2. Ten Factors

Based on information obtained from literature and in interviews with managers, the following 10 potential factors were chosen in relation to exerting an impact on the continued operation: (1) subsidies for initial investment; (2) subsidies for operation and maintenance; (3) farmer perception of RWHs; (4) doubts regarding rainwater quality; (5) availability of groundwater; (6) ownership; (7) location; (8) size of storage tank; (9) irrigation methods; and (10) technical problems. Of these 10 factors, factors (1)–(7) are non-technological and (8)–(10) are technological, which are presented in detail below. These impact factors are the most crucial to RWHs for agriculture irrigation in Beijing although they may not be the most representative ones.

(1) Subsidies for Initial Investment

Subsidies for initial investment can effectively assist in reducing the expenditure of farmers [8]. For example, the initial investment for a small plant with a capacity of 50 m^3 is approximately 27,000 RMB (approximately 4278 USD) including construction expenses. Small plants are generally constructed on farms with an area of 700 m^2. The total annual income of a small family farm in Beijing is, on average, approximately 10,000 RMB (approximately 1574 USD); thus, the initial investment is almost three times the annual income of the owners. When no subsidies are available, farmers have difficulty affording the large initial investment. Therefore, most RWHs in Beijing are provided with subsidies that cover approximately 50% to 100% of the initial investment, although some systems receive subsidies covering less than 50%.

(2) Subsidies for Operation and Maintenance

Several studies have been conducted on recovery of the operational and maintenance costs of RWH [22–24]. The systems need to have a sound maintenance system to enable sustainability and thus insufficient investment in maintenance obstructs the operation of these systems. Subsidies provided for operation and maintenance therefore assist with cost recovery, and may allow continuation of the plant's successful operation. In this study, RWHs managed privately did not receive subsidies for their operation and maintenance, whereas systems managed by state-owned farms or institutes can more readily obtain government subsidies.

(3) Farmer Perception of RWHs

To ensure the success of RWHs, it is important that farmers have a positive perception of the system [8,25]. Local farmers in the Beijing area obtain irrigation water by pumping groundwater. The use of rainwater is typically considered to be more complex than the use of groundwater, despite the fact that technology involved in RWHs is relatively simple. Some farmers in Beijing have a positive perception because rainwater supplements the irrigation water supply, but others consider it unnecessary for unscientific reasons, such the inconvenience of using different methods simultaneously.

(4) Doubts Regarding Rainwater Quality

Beijing is an industrialized megacity with 13 million residents and a large number of factories. The atmosphere is therefore contaminated by particles, heavy metals, and organic air pollutants [26]. Some users are skeptical about the quality of harvested rainwater due to contamination by air pollution [25] and are concerned that possible pollutants from the atmosphere enter the water and cause an absence of minerals therein. In addition, there is a lack of systematic information for publication whether rainwater in Beijing is suitable or not, although some experiments have been conducted. Such doubts concerning the rainwater quality result in a certain amount of resistance to using rainwater for irrigation.

(5) Availability of Groundwater

The availability of groundwater is regarded as a potential impact factor in this study, although it is often not discussed in the literature. At present, no clear policies exist relating to the charge for groundwater in Beijing. Therefore, the motivation of famers to use rainwater depends on the availability of groundwater and the cost of accessing it, which is determined by whether the location is mountainous or flat. The availability of groundwater varies according to the position of the plants studied in this paper. For example, two plants are both located in the Mi Yun district (plant 6 and 7, as shown in Figure 1) and the groundwater at plant 7 is sufficient, while groundwater at plant 6 is extremely limited.

Groundwater only emerges in certain wells in summer and the wells are dry in other seasons and thus these places are labeled as areas of water scarcity. In other areas it is easy to access groundwater through pumps, and thus resources are sufficient and the locations are considered to be areas with water sufficiency. Farmers in areas where water is scarce have a greater incentive to use alternative water resources than farmers in areas of water sufficiency.

(6) Ownership

Responsibility of the daily management and operation of a RWHs is determined by its ownership [22]. Some of the plants studied in this paper were managed privately and others were managed by state-owned institutes. Thus the incentives involved in managing and operating the systems were different. In general, the incentives for private-plant managers to operate RWHs are reduced irrigation costs or an increased income. However, the incentive for state-owned-plant managers is not dependent on additional costs or income, and therefore these plants are more likely to be continually operated even when they are not cost-effective.

(7) Location

Although location is rarely discussed in literature, in this study, it is considered highly relevant as it determines the depth of groundwater and the cost of operating a RWHs. The cost of obtaining groundwater can affect a farmer's incentive to use rainwater [27]. Seven out of the 10 plants studied were located in the northern area of Beijing, which is mostly mountainous, and three were located in the southern area, which is relatively flat and dry. Accordingly, the groundwater level in the north is lower than that in the south. Table 1 shows the depth of groundwater at the northern and southern plants. Farmers living in the northern area need to pump groundwater from deeper wells than farmers in the south. Consequently, higher costs are incurred when pumping water in the northern area, and thereby increasing farmers' motivation to take rainwater.

Table 1. Depth of groundwater at northern and southern plants (2007–2008).

Ground Water	P_1	P_2	P_3	P_4	P_5	P_6	P_7	P_8	P_9	P_{10}
Depth	28	28	28	32	32	30	30	13	13	20

Notes: Plants P_1–P_7 are located in the north and plants P_8–P_{10} in the south. Data source: Beijing water bulletin (2007–2008).

(8) Size of Storage Tank

The size of the storage tank is discussed in relation to the successful operation of RWHs [3,28]. The optimal storage size is dependent on precipitation, collection area, water consumption, water saving, and economic issues [28]. In addition, the size of the storage tank can determine the initial investment and operation and maintenance costs involved. For instance, although the operation and maintenance costs of a larger system are lower per unit than the costs of a small system, the total initial investment and operation and maintenance costs of a larger system are higher. In this study, the storage tanks were classified into three categories: small (50–100 m^3), medium (450–800 m^3), and large (1300–2000 m^3).

(9) Irrigation Methods

Irrigation methods can affect the operation of RWHs [24]. Such methods normally include flood and drip irrigation, where flood irrigation is the traditional method often used by farmers in Beijing and drip irrigation is currently promoted because it effectively reduces water consumption. However, the initial expenditure for drip irrigation is high and not all farmers can afford it. In addition, plants using drip irrigation have problems frequently with mud blockages because the diameter of water pipes used in drip irrigation is small enough for mud and rocks to become trapped. Compared with flood irrigation, drip irrigation requires higher maintenance cost.

(10) Technical Problems

In literature, technical problems are the most-mentioned factors limiting the use of RWHs [23,29], and these can be divided into problems that are severe and those that are less severe. Severe problems consist of, for example, replacement parts required for the plant, and less severe problems consist of parts that are broken and need repair. For the construction of a plant, technical problems can be divided into small and large problems. Large problems involve the implementation of a design that is unsuitable for successful plant operation; small problems involve parts included in a design that require alteration (for example, a shallow ditch being modified to a deep ditch to improve rainwater collection).

This study combines two types of classification and codifies the various technical problems. As shown in Table 2, a coding of 0 indicates that the facility problems are less severe but that the construction design is inappropriate; 1 indicates that the facility needs repair and that the construction design has a small problem; 2 means that some parts of the facility and the construction design require changing; and 3 means that the technical problems of the plant are extremely severe. The facility and its construction are usually designed and implemented by the same institute, and thus they usually work well together. However, when the construction design is inappropriate the facilities are therefore also inadequate for plant operation and thus it is rare to use a coding of 0. In addition, as RWHs in Beijing are at an early stage of implementation and use, many types of technical problems occur during their operation. All cases studied here had a history of problems in either the facility or related to the construction design.

Table 2. Classification of technical problems.

Problems	Big	Small
Less serious problems	0	1
Serious problems	3	2

2.3. Rough Set Analysis

The information regarding the selected RWHs is a mixture of qualitative and quantitative data. So, the qualitative information is needed to be transferred to quantitative data for further information classification and data mining. Firstly, data pertaining to the 10 factors were all codified as 1, 2, or 3 to establish a consistent database (Table 3). Secondly, as 10 plants were selected for this study, a data matrix was formed for data pertaining to scores for 10 factors at each plant (Table 4). As shown in Table 4, the coded values of the 10 factors for the 10 plants were different, and each factor is denoted by A_i ($i = 1, 2, 3, \dots 10$); the status of operation is denoted by D.

The last row of Tables 3 and 4 indicates the operation status of each plant. Based on interviews with plants managers, one of three statuses was possible concerning the operation of each plant: stopped, interrupted, or continuous. The status of "stopped" means that the plant was no longer functioning, and was represented by a score of 1. The status of "interrupted" means that the plant was operational but that it stopped irregularly, and was represented by a score of 2. The status of "continuous" means that the plant was operating successfully and continuously, and this was represented by a score of 3. In this way, these scores describe the situations of the 10 constructed RWHs studied in Beijing.

Table 3. Description and coded values of factors.

Factors	Description and Coded Value
A_1: Ownership	1: Private 2: State owned
A_2: Farme perception of RWHs	1: Negative 2: Positive
A_3: Doubts regarding rainwater quality	1: Yes 2: No
A_4: Location	1: North 2: South
A_5: Availability of groundwater	1: Sufficient 2: Scarce
A_6: Size of storage tank	1: Small 2: Middle 3: Large
A_7: Irrigation methods	1: Flood irrigation 2: Drip irrigation
A_8: Technical problems	1: Less serious and small problems 2: Serious and small problems 3: Serious and big problems
A_9: Subsidies for initial investment	1: 51%–100% of the initial investment 2: 0%–50% of the initial investment
A_{10}: Subsidies for operation and maintenance	1: Yes
	2: No
D: Status of each plant	1: Stopped 2: Interrupted 3: Continuous

Table 4. Codified data matrix of 10 factors and operational status of 10 plants.

Factors	P_1	P_2	P_3	P_4	P_5	P_6	P_7	P_8	P_9	P_{10}
A_1: Ownership	1	2	1	1	1	1	1	2	1	2
A_2: Famer perception of RWHs	1	2	2	2	1	2	2	1	1	1
A_3: Doubts regarding rainwater quality	1	2	2	1	2	2	2	1	1	1
A_4: Location	1	1	1	1	1	1	1	2	2	2
A_5: Availability of groundwater	1	1	1	1	1	2	1	1	1	1
A_6: Size of storage tank	1	3	3	2	2	1	1	2	1	2
A_7: Irrigation methods	2	2	1	1	2	1	1	2	2	1
A_8: Technical problems	2	3	2	2	1	3	2	2	1	1
A_9: Subsidies for initial investment	1	1	1	1	1	1	1	2	1	2
A_{10}: Subsidies for operation and maintenance	2	1	1	2	1	2	2	1	2	2
D: Status of each plant	1	2	2	1	2	3	2	1	1	1

As the operational status of each plant was assumed to be affected by 10 identified factors, these 10 factors (A_1, A_2, A_3 ... A_{10}) were regarded as "condition" attributes, and the statuses of the plants (D) were regarded as "decision" attributes. We aimed to determine the causal links between the condition attributes and decision attributes to identify factors determining the cessation or continued operation of the RWHs.

We assumed that the 10 plants (P_1, P_2, P_3 ... P_{10}) belonged to a set U, namely

$$U = \{P_1, P_2, P_3, P_4, P_5, P_6, P_7, P_8, P_9, P_{10}\}$$

The plants were scored by each attribute based on the information available (shown in Table 4). For example, for $A_1 = 2$ (ownership is "state-owned"), Plants P_2, P_8, and P_{10} had the same score; in other words, Plants P_2, P_8, and P_{10} are all state-owned systems. Hence, the set U was classified into subsets based on the scored value of each attribute, as follows:

$$U/A_1 = \{\{P_1, P_3, P_4, P_5, P_6, P_7, P_9\}, \{P_2, P_8, P_{10}\}\}$$

$$U/A_2 = \{\{P_1, P_5, P_8, P_9, P_{10}\}, \{P_2, P_3, P_4, P_6, P_7\}\}$$

$$U/A_3 = \{\{P_1, P_4, P_8, P_9, P_{10}\}, \{P_2, P_3, P_5, P_6, P_7\}\}$$

$$U/A_4 = \{\{P_1, P_2, P_3, P_4, P_5, P_6, P_7\}, \{P_8, P_9, P_{10}\}\}$$

$$U/A_5 = \{\{P_1, P_2, P_3, P_4, P_5, P_7, P_8, P_9, P_{10}\}, \{P_6\}\}$$

$$U/A_6 = \{\{P_1, P_6, P_7, P_9\}, \{P_4, P_5, P_8, P_{10}\}, \{P_2, P_3\}\}$$

$$U/A_7 = \{\{P_3, P_4, P_6, P_7, P_{10}\}, \{P_1, P_2, P_5, P_8, P_9\}\}$$

$$U/A_7 = \{\{P_1, P_2, P_3, P_4, P_5, P_6, P_7\}, \{P_8, P_9, P_{10}\}\}$$

$$U/A_8 = \{\{P_5, P_9, P_{10}\}, \{P_1, P_3, P_4, P_7, P_8\}, \{P_2, P_6\}\}$$

$$U/A_9 = \{\{P_1, P_2, P_3, P_4, P_5, P_6, P_7, P_9\}, \{P_8, P_{10}\}\}$$

$$U/A_{10} = \{\{P_2, P_3, P_5, P_8\}, \{P_1, P_4, P_6, P_7, P_9, P_{10}\}\}$$

In addition, the set U was classified into a subset in terms of the coded value of the decision attribute, as follows:

$$U/D = \{\{P_1, P_4, P_8, P_9, P_{10}\}, \{P_2, P_3, P_5, P_7\}, \{P_6\}\}$$

If we assume:

$$Y_1 = \{P_1, P_4, P_8, P_9, P_{10}\}; \ Y_2 = \{P_2, P_3, P_5, P_7\}; \ Y_3 = \{P_6\},$$

then

$$U/D = \{Y_1, Y_2, Y_3\}.$$

where the subset Y_1 represents the group of plants with stopped operation, Y_2 represents the group of plants with interrupted operation, and Y_3 represents the group of plants with continuous operation. The sets U/A_i (i = 1,2,3 ... 10) may contain the same subset as set U/D. If the subset Y_j (j = 1,2,3) of the set U/D can be determined in any of the sets U/A_i (i = 1,2,3 ... 10), a link between D and A_i can be identified. For example, a set U/A_m (m = 1 or 2 or 3 ... or 10) containing the subset Y_n (n = 1 or 2 or 3) indicates that the group of plants with an nth operational status can be characterized by A_m. In other words, the attribute A_m is the critical factor affecting the decision attribute D.

The equation for identifying these types of linkages is as follows:

$$(U/A_i) \cap (U/D) = Y_j \ (i = 1,2,3 \ldots 10, j = 1,2,3) \tag{1}$$

Using the aforementioned equations, this study then determined the following:

The term $j = 1 \rightarrow i = 3$, meaning that U/A_3 contains Y_1; The term $j = 2 \rightarrow i = 3$ and $i = 5$, meaning that U/A_3 and U/A_5 contain Y_2; The term $j = 3 \rightarrow i = 5$, meaning that U/A_5 contains Y_3.

These results indicated that only A_3 and A_5 are linked with the decision attribute D. Accordingly, the conditional causal links of an "if ... , then ... " type were derived (Table 5), which are known as "rules" in rough set theory, where a rule specifies the relationships between condition and decision attributes.

Table 5. Rules involved in the operation of RWHs.

Rule Number	If	Then
1	$A_3 = 1$	$D = 1$
2	$A_3 = 2$ and $A_5 = 1$	$D = 2$
3	$A_5 = 2$	$D = 3$

3. Results and Discussion

Although a considerable amount of money has been invested into RWHs in Beijing, these systems are not often used. This research identified three rules involved in their operation (Table 5) through rough set analysis. The first rule is that when farmers have doubts about the quality of rainwater, operation of the system is discontinued. The second rule is that although farmers have no doubts about the quality of rainwater, they still do not operate the plant continuously if it is possible to obtain sufficient groundwater. The third rule is that only when there is a shortage of groundwater, RWHs operated continuously and successfully. These results are summarized in Table 6. Rule 1 was supported by five plants (P_1, P_4, P_8, P_9, and P_{10}), Rule 2 by four plants (P_2, P_3, P_5, and P_7), and Rule 3 by one plant (P_6).

Table 6. Description of rules and plants concerned.

Rule	If	Then	Plants
1	Rainwater quality is doubted	Plant operation ceases	P_1, P_4, P_8, P_9, P_{10}
2	Rainwater quality is not doubted but groundwater source is sufficient	Plant operation interrupted	P_2, P_3, P_5, P_7
3	Groundwater source is not sufficient	Plant operation continues	P_6

This study also reveals that the decisive factors concerning whether or not RWHs are operated ("doubts about the rainwater quality" and "availability of groundwater") are both non-technological factors. In comparison with technological factors ("size of storage tank", "irrigation method", and "technical problems"), the two non-technological factors play a critical role in the decision to continue operation of RWHs. These results verify the importance of non-technological factors in the sustainable management of water resources.

To promote RWH for agricultural irrigation in Beijing, the government currently subsidizes and has introduced suitable technology [30]. These actions may facilitate the introduction of RWH. However, continuous operation of RWH depends on the confidence of users in the water quality and their motivation to use the rainwater. Some residents of Beijing believe that city rainwater contains a number of chemicals related to the severe air pollution so that it is unsuitable for agricultural irrigation. RWHs tend to fail when farmers doubt the quality collected rainwater (Rule 1). It is, thus, necessary to alleviate doubts about the quality of rainwater as a crucial first step in promoting the use of RWH. Farmers would use rainwater when they considered it safe for irrigation. Testing of rainwater quality should be performed, and the results should be revealed and explained to the public.

However, some farmers do not doubt the water quality, and there are no technical or financial problems involved, but as they can easily obtain groundwater, the operation may still not continue (Rule 2). According to our interviews and literature review, barriers to using groundwater in Beijing are extremely low [30,31]. Thus, sufficient groundwater resources and low barriers to obtaining the water reduce the motivation on RWHs. Conversely, a shortage of groundwater resources raises farmers' incentive to search for and use other sources of water for irrigation. When groundwater is scarce, RWH continue to operate successfully and continuously (Rule 3). Hence, increasing barriers to obtaining groundwater in water-sufficient areas is a crucial second step in promoting RWHs. Various measures, such as groundwater charges, prohibiting the pumping of new wells, and limiting the quantity of groundwater pumping, may be required [7]. It is interesting to note the importance of subjective perception, as in practice, farmers have little information about the pollution of rain water, but perceived pollution causes them to continue using groundwater.

4. Conclusions

The present study aimed to examine the effect of non-technological factors on sustainable RWHs management, through identifying critical drivers of the success or failure of RWHs for agriculture irrigation in Beijing. The method of rough set analysis, was applied to analyze partially qualitative and partially quantitative data collected on the functioning of RWH in the Beijing rural area.

Two factors, namely "doubts about rainwater quality" and "availability of groundwater," were determined as decisive factors for the decision to continue or stop using the RWHs. In other words, as long as farmers have doubts about rainwater quality or they consider that there is sufficient groundwater, the newly constructed RWHs will not continue to be operational in the long term. However, if there is a perceived groundwater shortage, farmers will operate the RWHs continuously. Therefore, to enable the sustainable and successful operation of RWHs in Beijing, it is necessary to determine how to improve confidence and motivation in using rainwater. Removing doubts about rainwater quality is an essential step in the promotion of RWHs in Beijing.

Through a case study of RWHs in Beijing, the present study demonstrated that non-technological factors are critical for sustainable water management. It is considered that non-technological factors should receive more attention from researchers and decision makers, as in this instance, they are significant factors in determining the successful, long term operation of this sustainable water-management system.

Acknowledgments: Acknowledgments: This research was funded by Natural Science Foundation of SZU (Grant No. 201428). We are grateful to the anonymous reviewers for their insightful comments.

Author Contributions: Author Contributions: The article was mainly written by Xiao Liang. Meine Pieter van Dijk provided many valuable comments.

Conflicts of Interest: Conflicts of Interest: The authors declare no conflict of interest.

References

1. Campisano, A.; Modica, C. Regional scale analysis for the design of storage tanks for domestic rainwater harvesting systems. *Water Sci. Technol.* **2012**, *66*, 1–8. [CrossRef] [PubMed]
2. Burns, M.J.; Letcher, T.D.; Duncan, H.P.; Hatt, B.E.; Ladson, A.R.; Walsh, C.J. The performance of rainwater tanks for stormwater retention and water supply at the household scale: An empirical study. *Hydrol. Processes* **2015**, *29*, 152–160. [CrossRef]
3. Mishra, A.; Adhikary, A.K.; Panda, S.N. Optimal size of auxiliary storage reservoir for rain water harvesting and better crop planning in a minor irrigation project. *Water Resour. Manag.* **2009**, *23*, 265–288. [CrossRef]
4. Woltersdorf, L.; Liehr, S.; Döll, P. Rainwater harvesting for small-holder horticulture in namibia: Design of garden variants and assessment of climate change impacts and adaptation. *Water (Switzerland)* **2015**, *7*, 1402–1421. [CrossRef]
5. Ghimire, S.R.; Johnston, J.M.; Ingwersen, W.W.; Hawkins, T.R. Life cycle assessment of domestic and agricultural rainwater harvesting systems. *Environ. Sci. Technol.* **2014**, *48*, 4069–4077. [CrossRef] [PubMed]
6. Karim, M.R.; Bashar, M.Z.I.; Imteaz, M.A. Reliability and economic analysis of urban rainwater harvesting in a megacity in bangladesh. *Resour. Conserv. Recycl.* **2015**, *104*, 61–67. [CrossRef]
7. Liang, X.; van Dijk, M.P. Economic and financial analysis on rainwater harvesting for agricultural irrigation in the rural areas of beijing. *Resour. Conserv. Recycl.* **2011**, *55*, 1100–1108. [CrossRef]
8. Zuo, J.; Liu, C.; Zheng, H. Cost-benefit analysis for urban rainwater harvesting in Beijing. *Water Int.* **2010**, *35*, 195–209.
9. Domènech, L.; Saurí, D. A comparative appraisal of the use of rainwater harvesting in single and multi-family buildings of the metropolitan area of barcelona (spain): Social experience, drinking water savings and economic costs. *J. Clean. Product.* **2011**, *19*, 598–608. [CrossRef]

10. Matos, C.; Bentes, I.; Santos, C.; Imteaz, M.; Pereira, S. Economic analysis of a rainwater harvesting system in a commercial building. *Water Resources Manag.* **2015**, *29*, 3971–3986. [CrossRef]
11. Baiyegunhi, L.J.S. Determinants of rainwater harvesting technology (rwht) adoption for home gardening in msinga, kwazulu-natal, south africa. *Water SA* **2015**, *41*, 33–40. [CrossRef]
12. Recha, C.W.; Mukopi, M.N.; Otieno, J.O. Socio-economic determinants of adoption of rainwater harvesting and conservation techniques in semi-arid tharaka sub-county, kenya. *Land Degrad. Dev.* **2015**, *26*, 765–773. [CrossRef]
13. He, X.F.; Cao, H.; Li, F.M. Econometric analysis of the determinants of adoption of rainwater harvesting and supplementary irrigation technology (rhsit) in the semiarid loess plateau of china. *Agric. Water Manag.* **2007**, *89*, 243–250. [CrossRef]
14. Zhang, Z.; Xu, Z. Rough set method to identify key factors affecting precipitation in lhasa. *Stoch. Environ. Res. Risk Assess.* **2009**, *23*, 1181–1186. [CrossRef]
15. Pawlak, Z. Rough sets. *Int. J. Comput. Inf. Sci.* **1982**, *11*, 341–356. [CrossRef]
16. Slowinski, R. *Intellligent Decison Support: Handbook of Applications and Advances of Rough Set Theory*; Kluwer Academic Publishers: Norwell, MA, USA, 1991.
17. Barbagallo, S.; Consoli, S.; Pappalardo, N.; Greco, S.; Zimbone, S.M. Discovering reservoir operating rules by a rough set approach. *Water Resour. Manag.* **2006**, *20*, 19–36. [CrossRef]
18. Nijkamp, P.; Van der Burch, M.; Vindigni, G. A comparative institutional evaluation of public-private partnerships in dutch urban land-use and revitalisation projects. *Urban. Stud.* **2002**, *39*, 1865–1880. [CrossRef]
19. Baycan-Levent, T.; Nijkamp, P. Planning and management of urban green spaces in europe: Comparative analysis. *J. Urban. Plan. Dev.* **2009**, *135*, 1–12. [CrossRef]
20. Wu, C.; Yue, Y.; Li, M.; Adjei, O. The rough set theory and applications. *Eng. Comput.* **2004**, *21*, 488–511. [CrossRef]
21. Nijkamp, P.; Rietveld, P.; Spierdijk, L. A meta-analytic comparison of determinants of public transport use: Methodology and application. *Environ. Plan. B: Plan. Des.* **2000**, *27*, 893–903. [CrossRef]
22. Mushtaq, S.; Dawe, D.; Hafeez, M. Economic evaluation of small multi-purpose ponds in the zhanghe irrigation system, China. *Agric. Water Manag.* **2007**, *91*, 61–70. [CrossRef]
23. Oweis, T.; Hachum, A. Water harvesting and supplemental irrigation for improved water productivity of dry farming systems in west asia and north Africa. *Agric. Water Manag.* **2006**, *80*, 57–73. [CrossRef]
24. Berthelot, P.B.; Robertson, C.A. A comparative study of the financial and economic viability of drip and overhead irrigation of sugarcane in mauritius. *Agric. Water Manag.* **1990**, *17*, 307–315. [CrossRef]
25. Song, J.; Han, M.; Kim, T.; Song, J. Ras a sustainable water supply option in banda aceh. *Desalination* **2009**, *248*, 233–240. [CrossRef]
26. Helmreich, B.; Horn, H. Opportunities in rainwater harvesting. *Desalination* **2009**, *248*, 118–124. [CrossRef]
27. Liang, X.; van Dijk, M.P. Optimal level of groundwater charge to promote rainwater usage for irrigation in rural beijing. *Water* **2011**, *3*, 1077–1091. [CrossRef]
28. Campisano, A.; Modica, C. Optimal sizing of storage tanks for domestic rainwater harvesting in sicily. *Resour. Conserv. Recycl.* **2012**, *63*, 9–16. [CrossRef]
29. Hatibu, N.; Mutabazi, K.; Senkondo, E.M.; Msangi, A.S.K. Economics of rainwater harvesting for crop enterprises in semi-arid areas of east africa. *Agric. Water Manag.* **2006**, *80*, 74–86. [CrossRef]
30. Wang, Y.; Wang, H. Sustainable use of water resources in agriculture in beijing: Problems and countermeasures. *Water Policy* **2005**, *7*, 345–357.
31. Yang, H.; Zhang, X.; Zehnder, A.J.B. Water scarcity, pricing mechanism and institutional reform in northern china irrigated agriculture. *Agric. Water Manag.* **2003**, *61*, 143–161. [CrossRef]

water

MDPI

Article

A Reliability Analysis of a Rainfall Harvesting System in Southern Italy

Lorena Liuzzo [1,*], Vincenza Notaro [2] and Gabriele Freni [1]

[1]   Facoltà di Ingegneria ed Architettura, Università degli Studi di Enna Kore, Enna 94100, Italy; gabriele.freni@unikore.it

[2]   Dipartimento di Ingegneria Civile Ambientale Aerospaziale e dei Materiali, Università degli Studi di Palermo, Palermo 90128, Italy; vincenza.notaro@unipa.it

*   Correspondence: lorena.liuzzo@unikore.it; Tel.: +39-0935-536439

Academic Editor: Ataur Rahman

Received: 30 September 2015; Accepted: 5 January 2016; Published: 8 January 2016

Abstract: Rainwater harvesting (RWH) may be an effective alternative water supply solution in regions affected by water scarcity. It has recently become a particularly important option in arid and semi-arid areas (like Mediterranean basins), mostly because of its many benefits and affordable costs. This study provides an analysis of the reliability of using a rainwater harvesting system to supply water for toilet flushing and garden irrigation purposes, with reference to a single-family home in a residential area of Sicily (Southern Italy). A flushing water demand pattern was evaluated using water consumption data collected from a sample of residential customers during an extended measurement campaign. A daily water balance simulation of the rainwater storage tank was performed, and the yield-after-spillage algorithm was used to define the tank release rule. The model's performance was evaluated using rainfall data from more than 100 different sites located throughout the Sicilian territory. This regional analysis provided annual reliability curves for the system as a function of mean annual precipitation, which have practical applications in this area of study. The uncertainty related to the regional model predictions was also assessed. A cost-benefit analysis highlighted that the implementation of a rainwater harvesting system in Sicily can provide environmental and economic advantages over traditional water supply methods. In particular, the regional analysis identified areas where the application of this system would be most effective.

Keywords: rainwater harvesting; flushing water demand; water balance simulation; reliability analysis

1. Introduction

Increasing water demand has led to water scarcity in many urban areas in the Mediterranean region. Indeed, population growth and the expansion of urban and industrialized areas has put great pressure on water resources. Climate change will intensify this pressure in some parts of the world, including the Mediterranean basin, Western United States and Southern Africa, resulting in a predicted decrease in water resources in the coming decades [1]. In this context, developing strategies and systems to identify alternative water resources will become critical, as will improving water resources management and planning. Water desalination and recycling processes, together with intermittent water supply, have long been the most common technologies used to cope with water scarcity in urban areas, while the benefits of collecting and using rainwater have largely been ignored [2,3]. Nevertheless, rainwater has historically been the primary source of water for potable and non-potable uses in locations where water supply systems have not yet been developed, and has traditionally been employed in a variety of ways in new settlements and isolated homes [4]. Because of their many environmental and economic advantages, rainwater harvesting (RWH) systems are currently receiving increased attention as alternative sources of drinking water, especially in semi-arid areas [5–7], but also in urban areas [8].

Generally, RWH systems involve three principal components: the catchment area, the collection device and the conveyance system. Rainwater is commonly collected from rooftops, courtyards or other compacted or treated surfaces before being filtered and collected in storage tanks to be used. RWH has many benefits. First, it requires simple and inexpensive technologies that are easy to install and maintain. Because of their simplicity, RWH systems can be expanded, reconfigured or relocated to meet each household's needs. RWH also has important economic advantages for consumers because it reduces the amount of water purchased from public systems. Moreover, the possibility of having an alternative water supply reduces pressure on aquifers and surface water sources. For these reasons, the integration of RWH systems into buildings is an effective way to minimize the use of treated water for non-potable tasks and supply drinking water in places where water is scarce.

While RWH has numerous benefits, there are some disadvantages, particularly related to the limits of its supply and the reliability of rainfall (both in terms of spatial and temporal distribution). For these reasons, RWH systems cannot supply water for all domestic uses and are unlikely to make the households independent of the conventional water supply system. To achieve water self-sufficiency, multiple technologies must be employed. Nevertheless, the acquisition and use of rainwater through RWH can provide a considerable amount of water and ensure substantial financial savings to households.

The quantity and quality of collected rainwater depends on geographic location, local climate characteristics, the presence of anthropic activities in the area and storage tank volume. In general, rainwater is relatively clean, has low hardness and a quasi-neutral pH, and is free of sodium [9]. Runoff from rooftops is often considered unpolluted [10] or at least is of relatively good quality compared with runoff from surface catchments [11]. However, there is still disagreement about the quality of rooftop runoff, ranging from good or acceptable [12,13] to contaminated [14,15], depending on the roofing material, environmental conditions and atmospheric pollution. Subject to basic treatments such as filtration and/or chlorination, as necessary, collected rainwater can be utilized for different non-potable uses, including toilet flushing, washing machine use and garden irrigation (or any other use that does not require high-quality water). Different studies have highlighted the benefits of using harvested rainwater for toilet flushing [16,17]. Zhang *et al.* [18] observe that harvesting all roof runoff for use in toilet flushing can reduce water consumption in residential buildings by about 25%.

The performance and design of RWH systems has been investigated using different approaches, including water balance simulation analyses and mass curve analyses [19–21], probabilistic methods [22] and economic optimization [3]. The results indicate that the storage capacity of tanks cannot be standardized but is considerably influenced by local rainfall, catchment surface characteristics and the number of people in the household.

Several studies have explored the implementation of RWH systems in response to growing water demand in Africa [7,23,24], Asia [25–27], USA. [17–28] and Australia [18–29]. Additional studies on RWH systems have been carried out in the Mediterranean region as well. In Greece, Sazakli *et al.* [30] analyzed the quality and utilization of rainwater for domestic and drinking purposes. In Spain, Farreny *et al.* [8] analyzed the cost-efficiency of an RWH system in a high-density social housing neighbourhood comprised of multi-storey buildings. In Southern Italy, Campisano and Modica [31] defined a dimensionless methodology to derive the optimal design of RWH systems for domestic use. This methodology was based on the results of daily water balance simulations carried out for 17 rainfall gauging stations.

The present study investigated the performance of a proposed RWH tank for a model single-family home in a residential area. Performance was tested for varying levels of annual precipitation using data from over 100 different sites in Sicily. The application of the yield-after-spillage algorithm enabled an evaluation of site-specific system efficiency. Performance was assessed for three tank sizes (10, 15 and 20 m^3) and three uses of the collected rainwater: toilet flushing, garden irrigation and both uses combined. Simulations were run using data from 2002 to 2004. The researchers analyzed water consumption data recorded from single-family homes in Palermo (Northwestern Sicily) during the selected time period to define a temporal pattern for flushing water demand. Water demand for garden irrigation was defined using recorded mean monthly evapotranspiration rates. Once the system's performance was evaluated for the entire study area, its reliability was analyzed as a function of mean annual precipitation to determine mathematical expressions that have regional validity and could be practically applied. A data resampling procedure was applied to evaluate uncertainty related to the regional model previsions. Finally, a cost-benefit analysis was performed in order to estimate the payback period on the capital cost for the RWH system installation.

The study highlights the limits and benefits related to the application of RWH systems in the area of study. In particular, the regional analysis allowed researchers to identify areas in which the installation of the selected RWH system would be most effective and for which rainwater uses.

2. Materials and Methods

2.1. Dataset

The present analysis uses data from Sicily, one of the 20 administrative regions in Italy, as a case study for a selected rainwater harvesting system. Sicily is an island of approximately 25,700 km^2 located in Southern Italy and is characterized by a Mediterranean climate (mild winters and hot, generally dry summers). The total annual rainfall in this area ranges from 400 mm/year at lower elevations to 1300 mm/year at higher elevations. Figure 1 shows the spatial distribution of mean annual precipitation over the 1981–2012 period in Sicily.

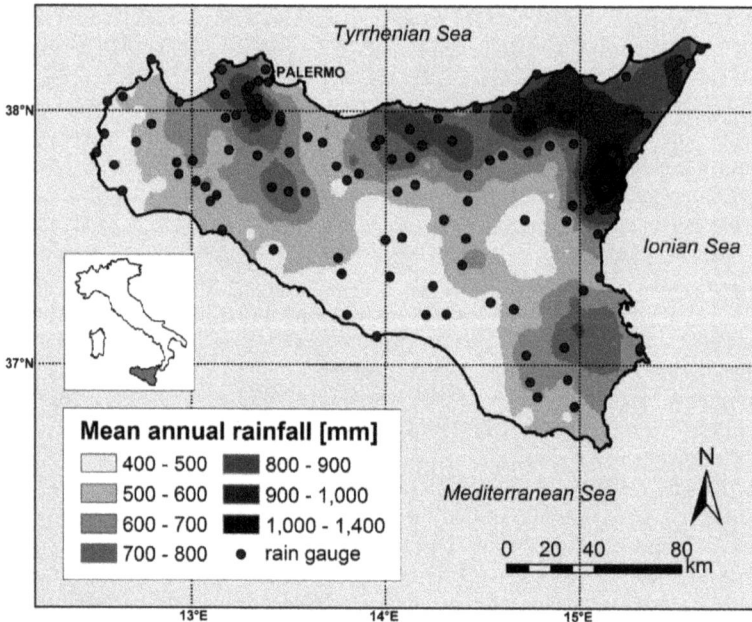

Figure 1. Spatial distribution of mean annual rainfall for the 1981–2012 period and locations of rain gauges.

Figure 2 illustrates the RWH system analyzed and provides a diagram of the different surface materials and their areas (m^2) onsite. The water catchment surfaces of the model home include the home's rooftop and the courtyard, for a total catchment area of 180 m^2 (100 m^2 of rooftop and 80 m^2 of courtyard and pedestrian areas). In this simulation, rainfall is collected from these surfaces and stored in a rainwater tank for two non-potable uses: toilet flushing and garden irrigation.

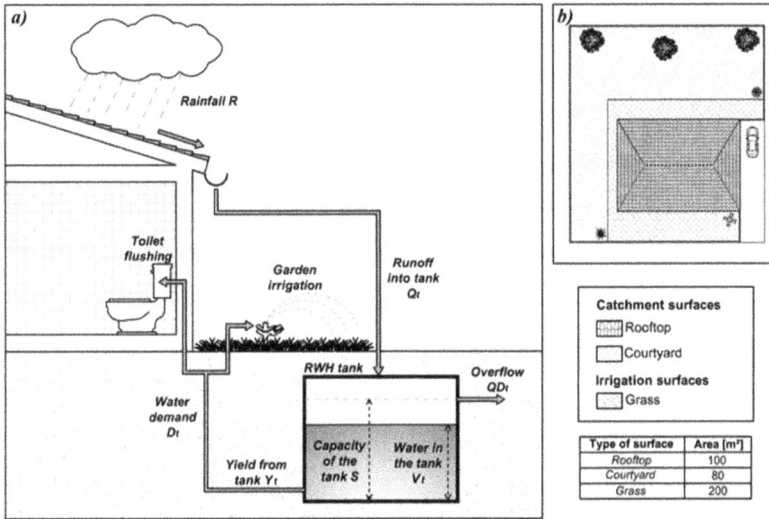

Figure 2. (**a**) Scheme of the RWH system; and (**b**) layout of the single-family house with the representation of the different surfaces.

The implementation of an RWH system requires an evaluation of the water balance, for which rainfall represents inflow and water demand for toilet flushing or for garden irrigation is the outflow. In the present study, rainfall volumes were calculated using the daily rainfall series recorded from 111 rain gauges over the 2002–2004 period (Figure 1). Rainfall data were provided by the *Osservatorio delle Acque-Agenzia Regionale per i Rifiuti e le Acque* (OA-ARRA) of Sicily. This period was chosen because a large number of the evenly distributed rain gauges that monitor rainfall throughout the Sicilian territory worked continuously during the entire period. This historical rainfall series is representative of the regional climate both in terms of annual and monthly mean values.

Water demand for flushing was calculated as the number of daily flushes per capita, which was obtained by analyzing water consumption data collected at a high temporal resolution from four-person single-family homes in Palermo (Northwestern Sicily) during a two-year measurement campaign. Water demand for garden irrigation was evaluated by estimating the mean monthly reference evapotranspiration. Historical temperature data obtained from the OA-ARRA for the 1981–2012 period were used for this calculation.

2.2. Inflow to the RWH Tank

The modelled rainwater tank is filled exclusively using rainfall volumes from a building's rooftop, courtyard and pedestrian areas. Assuming constant rainfall within each time step *t*, the rainwater volume can be calculated as follows:

$$Q_t = \phi \cdot A_{TOT} \cdot R_t = A \cdot R_t \tag{1}$$

where Q_t is the inflow volume supplied to the tank at time step *t* (m^3), φ is the runoff coefficient depending on water loss (dimensionless), R_t is the rainfall at time *t* (m), A_{TOT} is the total catchment surface area (m^2), and A is the effective impervious surface area (m^2). Evaporation losses from the tank are neglected. In this study, φ was set equal to 0.9 [32].

The stormwater quality of the initial discharge from the roof surface was of poor quality due to an accumulation of dust, sediments, bird and animal droppings, and leaves and debris from the surrounding areas [33], all of which were accumulated during the dry periods and washed off at the beginning of the next rain. The first flush is defined as the initial period of a rainwater runoff where a pollutant concentration is considerably higher than during later periods [34]. Depending on the specific site characteristics, type of contaminant and final use of the water, the literature provides different values of the amount of water that has to be diverted to ensure an adequate water quality. Yaziz *et al.* [35] and Coombes [36] reported that subtracting the first 0.33 mm of rainfall from the total daily rainfall as the first flush would significantly improve roof water quality. Following this recommendation, all the daily water balance simulations have been performed subtracting the first flush of 0.33 mm from the daily rainfall series.

2.3. Water Demand for Toilet Flushing

Estimating the average number of daily flushes per capita could be considered satisfactory to accurately model daily water demand for toilet flushing; however, these observations may not be universally applicable to all rainwater collection systems. Therefore, demand patterns with significant daily variations may require more precise modeling.

The water balance at the rainwater tank in the present study was evaluated at daily scale. The toilet flushing demand pattern was determined by analyzing water consumption data collected during a monitoring campaign of seven dwellings located in Palermo (Northwestern Sicily) throughout 2002–2004 (Figure 3).

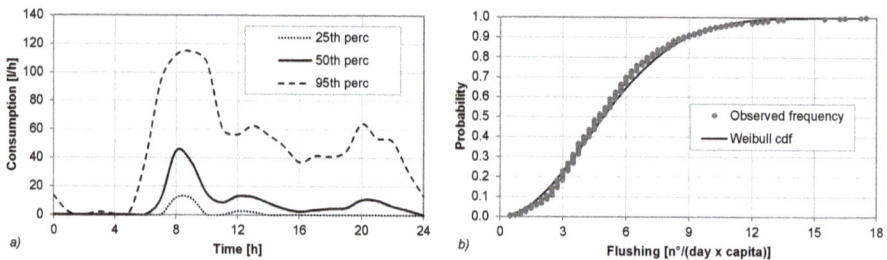

Figure 3. (**a**) Water demand percentiles of recorded data for Dwelling 6; and (**b**) Weibull cumulative distribution function CDF fitting the cumulated frequency of the number of daily flushing per capita for Dwelling 6.

The customers that participated in the consumption monitoring program had the following characteristics: families with at least two members; family members ranging in age from 4 to 70 years; negligible outdoor consumption; and interest in participating. Each monitored dwelling had a toilet WC flush tank with a volume of 9–10 L (the usual volume for a WC flush tank in Italy) and a bowl filling time ranging between 0.95 and 1 min.

An instrument package, including a Class C multi-jet water meter and a data logger, was installed on the service line of each of the seven dwellings downstream of the revenue water meter to monitor domestic water use. The two devices were coupled by means of an impulse sensor. When cumulative volume consumed equaled 0.5 L, the sensor transmitted a signal to the data logger. A common faucet is characterized by flows in the range 6–12 L/min, and the meter was able to disclose consumption pulses longer than or equal to 5 s (in the worst case) or equal to 2.5 s (in the best case), allowing researchers to separate out toilet flushing data from other uses. In any case, if small pulses were not identifiable, their volume was aggregated into the next consumption pulse. Cumulative volumes of more than 0.5 L were recorded in a text file containing six fields (*i.e.*, day, month, year, hour, minute and second). Water demand data were collected periodically by connecting the data logger to a laptop. The monitoring period was approximately one year for five dwellings, less than one year for two dwellings, and more than two years for one dwelling (Table 1). The monitoring period was long enough to identify weekly, monthly and seasonal toilet flushing patterns and was clear enough to identify user presence at home.

Once the data were acquired, according to the procedure proposed by Campisano and Modica [31], as first step of the analysis, the number of daily flushing was evaluated for each dwelling and monitoring day. To this purpose, the water consumption data were filtered to identify data points where use ranged from 9 to 10 L over a period of one minute. Knowing the filling time of the WC flush tank was important to exclude consumption data with the same volume but linked to other uses. In the absence of more specific information, the number of daily flushes per capita was then calculated for all monitored days as the number of flushes per day divided by the average number of users present, or the number of family members in each monitored household.

Table 1. Results of statistical analysis carried out on water consumption data collected for seven dwellings located in Palermo and monitored throughout 2002–2004.

Dwelling	$n°$ Persons	Monitoring Days	Average Flushings/(Day·Capita)	RMSE	CDF	λ	κ	K-S Test $D_{0.05}$
1	3	334	5.73	2.925	Weibull	6.66	2.234	0.071
2	4	359	5.77	2.951	Weibull	6.57	2.094	0.068
3	2	317	4.79	2.978	Weibull	5.77	1.912	0.060
4	3	237	4.62	2.974	Weibull	5.34	1.654	0.065
5	2	212	6.46	2.883	Weibull	7.31	2.410	0.077
6	4	637	5.12	2.798	Weibull	5.90	2.020	0.022
7	3	320	4.75	2.980	Weibull	5.35	1.674	0.059

The average number of daily flushes per capita for each monitored dwelling is reported in Table 1, along with the associated Root Mean Square Error (RMSE). These values ranged from 4.62 (Dwelling 4) to 6.46 (Dwelling 5) daily flushes per capita. The related RMSE was approximately 2.9 for five dwellings, 2.883 for one dwelling, and 2.798 for the final dwelling. These results are similar to those reported in previous studies available in the literature [31–37]. The number of daily flushes per capita were then statistically analyzed to identify a well-fitting probability distribution function. Several probability distribution functions were investigated, including the Normal, Poisson, Weibull, Exponential, *etc.* All monitored dwellings revealed similar statistical behaviors; the Weibull distribution function fit the observed data best. This was confirmed using the Kolmogorov-Smirnov statistical test (confidence level equal to 0.05). Table 1 also reports data for the two parameters λ and κ of the related Weibull distribution function together with the results of the Kolmogorov-Smirnov test for each dwelling.

An analysis of the processed data revealed that Dwelling 6 was representative of all monitored dwellings, with an average number of flushes per capita per day equal to 5.12 and a minor RMSE value equal to 2.798. Moreover, this household was continuously monitored for the longest period of time (around two years). Therefore, the subsequent RWH analysis uses Dwelling 6 to define the water demand pattern for toilet flushing in Sicily. Figure 3a shows the percentiles (25th, 50th and 95th) of the water demand data collected for Dwelling 6 during the monitoring campaign. Figure 3b shows the Weibull cumulative distribution function CDF fitting the cumulated frequency of the obtained per capita flushes for Dwelling 6.

To generalize the results to other similar users, 365 random points were sampled from this CDF to construct a daily pattern for an entire year of toilet flushes per capita. Finally, the series of daily household toilet flushes was computed by multiplying the number of flushes derived in the previous step by a selected number of users at home during the day.

2.4. Water Demand for Garden Irrigation

The frequency of irrigation depends on the type of grass, soil properties, and climatic conditions at the examined site. To evaluate the water demand for garden irrigation, it was assumed that the garden area (200 m^2) of the modelled single-family house was planted with turfgrass. To evaluate water demand, the mean monthly reference evapotranspiration ET_0 value was calculated for the area of study using the Thornthwaite formula [38]. ET_0 approximates water use for an irrigated grass pasture; therefore, water use for turfgrasses was estimated using a correlation factor, the crop coefficient K_c, as follows:

$$ET = ET_0 \cdot K_c \tag{2}$$

where ET is the actual evapotranspiration in mm/day. Turfgrass K_c values fluctuate slightly during the season based on the percentage of plant cover, growth rate, root growth, stage of plant development and management practices. In this study, K_c was set equal to 0.85 [39].

Once the amount of water to be provided was determined, the frequency of irrigation was defined based on practical considerations and previous literature. Optimum irrigation frequency depends on site, plant species, climatic conditions and soil types. Some studies (e.g., [40,41]) have highlighted that deep and infrequent irrigation promotes plant tolerance to drought stress. In a hot, humid region of the US, Jordan *et al.* [42] showed that irrigating every 4 days produced a larger and deeper root system. Moreover, irrigation scheduling is a process that requires knowledge of the irrigation system's characteristics, such as application rate and distribution uniformity. Watering frequency will vary from site to site and should be determined by the appearance of the turf. During peak water demand, turfgrass irrigation should occur every two or three days depending on the soil texture and root depth. For extremely arid climates, and depending on the type of turfgrass, the irrigation interval should be daily; but, during the early spring and in fall and winter, the frequency or irrigation interval may be stretched to every five to seven days [43]. Marchione [44] investigated the effects of different irrigation regimes on turfgrasses in Southern Italy and showed that, in a Mediterranean climate characterized by low rainfall and high evapotranspiration rates during summer, irrigation regimes equal to 75% of the water deficit are not adequate to maintain an acceptable turf quality.

The need for additional information to define the optimal irrigation frequency for turfgrass required to make some assumptions in this study. Specifically, it was assumed that the garden was planted with a turfgrass more resistant to warm climates than other species, such as *Zoysia Japonica Compadre*. It was also assumed that the garden was only irrigated every 3 days during April, May and September, and on alternate days from June to August. Table 2 summarizes the potential and actual daily evapotranspiration and the irrigation frequency for each month the garden was irrigated with harvested rainwater.

Table 2. Potential and actual evapotranspiration (mm/day) and the irrigation frequency for each month of garden irrigation with harvested rainwater.

Month	Evapotranspiration (mm/day)		Irrigation Frequency
	Reference	Actual	
April	1.5	1.3	every 3 days
May	2.4	2.0	every 3 days
June	3.5	3.0	alternate days
July	4.3	3.7	alternate days
August	4.5	3.8	alternate days
September	3.5	3.0	every 3 days

2.5. Water Balance Simulation

Different models can be used to predict the performance of RWH systems [45,46]. Often simple mass balance approaches based on annual precipitation volumes are used. However, these procedures do not ensure a proper level of accuracy in sizing RWH systems. Behavioural models are also frequently applied because they allow a more detailed design and are relatively simple to develop, although Ward *et al.* [46] showed that they usually underestimate the need for storage tank capacity compared with simple mass balance simulations.

In a behavioral model, the changes in the storage content of a finite reservoir are computed using the water balance equation. In this model, water fluxes consist of runoff into a tank (inflow), overflow from the tank and the yield extracted from the tank; demand is met in each operating period to the extent that storage is available.

The algorithm for the model relies on a yield-after-spillage (YAS) operating rule [47]:

$$Q_{D_t} = \max \begin{cases} V_{t-1} + A \cdot R_t - S \\ 0 \end{cases} \tag{3}$$

$$Y_t = \min \begin{cases} D_t \\ V_{t-1} \end{cases} \tag{4}$$

$$V_t = \min \begin{cases} V_{t-1} + A \cdot R_t - Y_t \\ S - Y_t \end{cases} \tag{5}$$

where, Q_{Dt} (m^3) is the volume discharged as overflow from the storage tank at time step t, V_t (m^3) is the volume stored at time step t, Y_t (m^3) is the yield of rainwater from the storage tank at time step t, D_t (m^3) is the toilet and grass irrigation water demand at time step t, and S (m^3) is the tank storage capacity.

The performance of RWH systems is generally described in terms of volumetric reliability, expressed as the total actual rainwater supply over water demand, R_v:

$$R_V = \frac{\sum\limits_{t=1}^{T} Y_t}{\sum\limits_{t=1}^{T} D_t} \cdot 100 \tag{6}$$

where T is the total time period under consideration and R_t is the overall water savings that can be achieved by harvesting and using rainwater. Equation (6) provides a measure of how much water has been conserved in comparison to the overall demand, and is also referred to as water saving efficiency [45].

3. Results and Discussion

3.1. Evaluation of Daily Reliability

The historical rainfall series recorded at 111 rain gauges during the 2002–2004 period were used to evaluate the performance of the RWH system in Figure 2. First of all, a preliminary analysis was carried out in order to examine the effect of the tank capacity S on the daily reliability R_V and to identify the tank capacity providing the most feasible value of the average daily R_V for each site in Sicily (assuming the same system configuration in terms of catchment surface).

Several tank capacities S in the range 1–30 m^3 were considered. Water balance simulations were performed at daily scale, thus accounting for the effect of extreme rainfall of 24 h duration and dry spells on the RWH system. Namely, for any tank size, the daily average R_V of each site was computed on the entire analysis period. Then, the related percentiles values were estimated. Results are summarized in the box-whisker graphs in Figure 4.

Focusing on the median line (50th percentile), the average daily R_V grows with tank capacity: For S ranging between 1 and 30 m^3, R_V varies in the range from 43% to 94% for toilet flushing use; this rise is steeper for irrigation use, specifically from 31% to 95%, while it is moderate for the combined use (R_V ranging from 39% to 80%).

Regarding toilet flushing use, when S is equal to 10 m^3, the RWH system reliability is higher than 80% and is equal to 92% for a capacity of 20 m^3. Further increases of S produce a slight improvement of R_V, with an achievable maximum value equal to 94%. For irrigation use, the median line shows an higher dependence of R_V on S. The system is able to provide an R_V value equal to 95% in more than 50% of the analyzed sites when a capacity of 30 m^3 is accounted. For this use the temporal shift between the rainwater demand for irrigation (higher during summer months) and rainfall amounts (lower during summer months) highly affects RWH system performance: Higher tank capacities permit the storing of greater rainwater volumes in winter in order to satisfy irrigation demand in summer. This effect is mitigated if combined use is considered because, in this case, the rainwater demand is widespread throughout the entire year. Indeed, the average daily R_V slightly increases for capacities higher than 10 m^3.

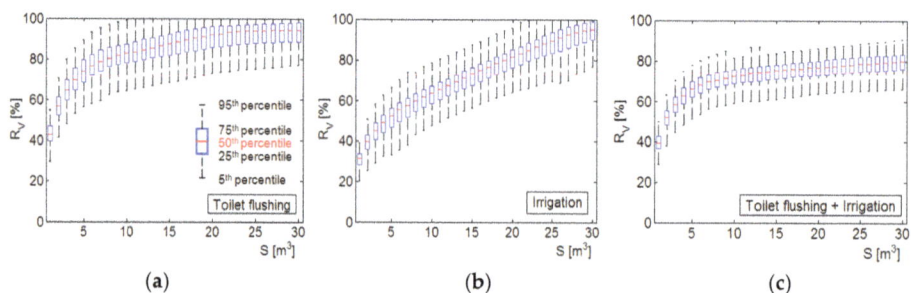

Figure 4. Box-whisker graphs of the daily reliability R_V vs. tank capacity S for different rainwater uses. (a) toilet flushing use; (b) irrigation use; (c) toilet flushing and irrigation use.

In order to assess the uncertainty linked to the R_V appraisal for each site, the average width of the R_V percentiles band (shown in Figure 4) was computed. Regarding the 25th and 75th percentile band, the average width values are equal to 19.8%, 8.8% and 7.1% for toilet flushing, irrigation and combined use, respectively. The average uncertainty regarding the 5th and 95th percentile band is 19.2% for combined use and about 24.5% for toilet flushing and irrigation. The reduced variability of R_V values among the analyzed sites for combined use highlights that rainwater demand represents a limiting factor to the achievement of higher RWH system performance in all the analyzed sites.

The performance improvement of RWH system in terms of R_V is moderate and not advantageous for tank capacity greater than 20 m^3 for toilet flushing and combined uses. Tank capacities higher than 20 m^3 may provide a significant improvement for irrigation use, but could be less economically feasible for a residential household (see Section 3.5 *Cost-benefit analysis*). Therefore, after this preliminary analysis, the performance of the RWH system were investigated focusing on three different capacities: 10, 15 and 20 m^3.

In order to analyze the effect of the temporal aggregation of the daily water balance output on R_V , the system performance was evaluated, for each site, at annual and monthly scales according to Equation (6). The following sections illustrate the obtained results.

3.2. Analysis of Annual Reliability

The annual reliability of the RWH system for each site of the studied area was assessed as average of the annual R_V values related to the three years chosen as the analysis period.

Figure 5 shows the spatial distribution of the annual reliability values over the study area. The use of the RWH system for toilet flushing provided the highest mean annual R_V values. The amount of water needed for toilet flushing for a family of four is approximately 80 m^3 per year. In the northwestern part of the island, where the mean annual precipitation ranges from 600 to 1000 mm, the performance of the system reached R_V values close to 100%, meaning that, in this area, the demand of water for toilet flushing can be completely satisfied by the water stored in an RWH system with a tank volume of just 10 m^3. Reliability was lower in sites located along the Mediterranean coast, where the mean annual precipitation ranges from 400 to 600 mm. In this zone, a 20 m^3 storage capacity was able to ensure reliability values up to 80%. A 10 m^3 RWH tank appears sufficient to ensure adequate R_V values in most of the area of study, while a larger capacity is required in the driest areas of the island. Conversely, a 10 m^3 storage capacity is not enough to meet the water demand for garden irrigation. Figure 5 shows that the use of an RWH system for garden irrigation results in poor performance. Specifically, for $S = 10$ m^3, the mean annual R_V was approximately 55%.

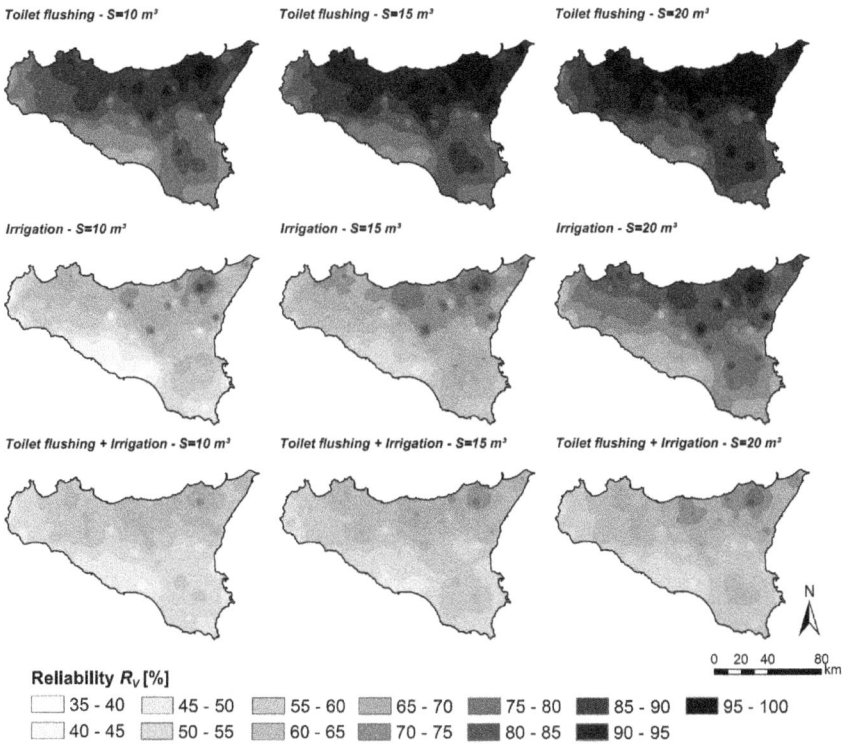

Figure 5. Spatial distribution of mean annual reliability R_V (%) for different rainwater uses and for S equal to 10, 15 and 20 m^3.

A wide area along the southern coast had R_V values that ranged from 35% to 45%. Therefore, a 10 m^3 storage capacity is not able to meet half of the annual water demand for garden irrigation. For this use, a 15 m^3 storage capacity increased reliability just 5% (R_V ranging from 45% to 50%). The use of a 20 m^3 tank was able to ensure good performance only in the northern part of the island, where the annual reliability of the system reached 80%; in the South, R_V ranged between 60% and 70%. To completely meet the water demand for garden irrigation, higher volumes of harvested rainwater are required. The mean annual demand for irrigation water is approximately 45 m^3; however, unlike the water demand for toilet flushing, which is homogeneously distributed over the year, irrigation demand is concentrated in spring and summer and has a peak in August. This temporal pattern deeply affects the performance of RWH systems because rainfall is scarce in Sicily during summer months, when increased evapotranspiration rates result in greater water demands for irrigation.

In the combined use case, the tank volumes considered in this analysis were not sufficient to ensure adequate system performances. The maps show that, when S is equal to 10 m^3, the average R_V was approximately 50%. Increased storage capacity up to 20 m^3 provided a slight increase in reliability, mainly in the northeastern part of the island, where the mean annual precipitation reaches 1,000 mm. Therefore, when limited rooftop and courtyard areas are available, the increase in storage volume is not enough to ensure the good performance of the RWH system, especially when rainwater must fill multiple needs with different temporal demand patterns, such as toilet flushing and garden irrigation. Furthermore, the increase in costs related to the installation of a larger storage tank makes the use of an RWH system less advantageous as capacity requirements increase.

3.3. Analysis of Monthly Reliability

To analyze the monthly variability of the RWH system's reliability, a separate analysis was performed for a particular location. The site selected for this analysis was Palermo, located on the northwestern coast of the island, where consumption data for toilet flushing were measured and analyzed. For S = 20 m^3, Figure 6a,d,g show plots of mean monthly demand, rainfall volumes and yield over the simulation period, as well as the corresponding monthly variation in reliability R_V when rainwater is used to flush toilets.

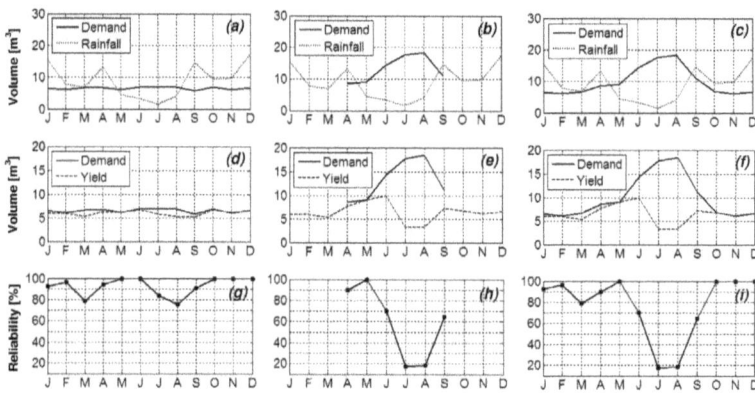

Figure 6. Monthly water demand, rainfall volume and yield, and monthly variation of system reliability for toilet flushing (**a,d,g**); garden irrigation (**b,e,h**); and both uses (**c,f,i**).

In Figure 6a, water demand and rainfall volumes are compared. Water demand for toilet flushing is clearly unaffected by monthly and seasonal variations, and shows only slight differences from month to month (on the order of 1 m^3), while rainfall volumes are affected by an evident seasonal pattern, with the lowest values occurring during summer months and the minimum value occurring in July. Figure 6d shows water demand and yield. When demand and yield overlap or the yield exceed the demand, the RWH system is able to completely meet the water demand for toilet flushing, ensuring a reliability of 100% (Figure 6g). Monthly R_V varies between 74% (in August) and 100% (in May, June, October, November and December).

In the case of garden irrigation (Figure 6b,e,h), the RWH system must provide water only during the period from April to September. Water demand is highest during summer months (Figure 5b), when temperatures are higher and evapotranspiration increases. Figure 6e shows that the demand exceeds the yield in June, July, August and September. This accounts for low monthly R_V values, especially in August when R_V equals 20% (Figure 6h), and means that a significant volume of water would need to be collected from other sources when rainwater is unavailable from the tank. For the examined site and the considered system, the use of rainwater for garden irrigation appears disadvantageous during summer months because the RWH system is not able to provide high levels of water savings compared to the costs incurred for system installation and maintenance. Because the water demand volumes are higher than the maximum capacity of the tank (20 m³), the poor performance of the system highlights the need to accumulate more rainwater during rainfall events by increasing the area of collection surfaces.

Figure 6c,f,i shows the results of the RWH system under the combined use scenario. The total water demand is the sum of monthly water volume required for toilet flushing and monthly water volume needed for garden irrigation (Figure 6c). The demand for irrigation is much higher than that for toilet flushing, as shown by the consistent increase in total water demand during the summer months. However, the water collected during the winter, spring and autumn months ensures adequate yields to meet the water demand for toilet flushing, reaching R_V levels up to 100% (Figure 6f). The performance of the RWH system clearly declines during the summer when the collected water is not enough to meet the higher demand for garden irrigation, resulting in a significant decrease in monthly R_V (Figure 6i).

3.4. Regional Reliability Curves and Related Uncertainty

The relationship between annual reliability and mean annual precipitation was investigated to define equations for a system analogous to the one analyzed here (for S equal to 10, 15 and 20 m³) and valid at the regional scale. The goal of these equations is to provide a reliability R_V that an RWH system can attain at an annual scale for each value of mean annual precipitation P and the uncertainty related to its estimation. Starting from simulation results previously shown, the points (P, R_V) were interpolated according to the following procedure:

- From the original dataset of annual reliabilities of the RWH system, which were obtained by applying the YAS algorithm to the 111 sites distributed over the Sicilian territory for the 2002–2004 period, 10,000 sub-datasets were extracted, in which 30% of points were randomly excluded to investigate the uncertainty affecting the results related to the selected sites;
- for each sub-dataset the interpolation curve was estimated;
- for each value of P, the 5th, 50th and 95th percentiles were obtained from the interpolation curves. The interpolation curve obtained for the 50th percentile represents the relationship between P and R_V, while the uncertainty related to the estimation of R_V as a function of P is given by the width of the interpolation curves for the 5th and 95th percentiles, respectively.

For each rainwater use and each value of S, Figure 7 shows the interpolation curves and the resulting uncertainty bands (dotted lines) obtained by interpolating the 5th and 95th percentiles. Table 3 shows the equation of the curves and the uncertainty bands. In general, reliability increases with mean annual precipitation and tank size. For the same values of P, the highest reliability can be obtained using the harvested rainwater only for toilet flushing. In this case, the RWH system is able to ensure an annual R_V that varies from 80% and 100% in locations characterized by a mean annual precipitation ranging from 600 to 1000 mm. According to these results, the installation of an RWH tank is particularly effective on the northeastern part of the island (as shown in Figure 5).

In terms of rainwater use for garden irrigation, when $S = 10$ m³ R_V does not reach 100% even at the sites with the highest mean annual precipitation values. Garden irrigation requires a storage of at least 20 m³ to obtain higher values of R_V; however, these values remain under 100%. The curves illustrate that the RWH system's performance declines if the rainwater is intended for the dual uses of toilet flushing and garden irrigation.

For every use, the evaluation of the system's reliability is affected by a lower level of uncertainty corresponding to a mean annual precipitation in the range from 600 to 1000 mm, as shown by the smaller width of the band. R_V values that exceed 100% indicate that the installation of an RWH system can completely meet the water demand and supply additional volume, which could be allocated to other uses. This occurs where the mean annual precipitation is greater than 1400, 1200 and 1100 mm when S equals 10, 15 and 20 m^3, respectively. However, the uncertainty related to higher values of P is greater than that related to the range 600–1000 mm, as shown by the increased width of the band of uncertainty.

Table 3. Equations of interpolating curves of 5th, 50th and 95th percentiles for each rainwater use and tank volume.

Rainwater Use	Tank Volume(m³)	P-R_V Curve	Uncertainty Bands	
		50th Percentile	5th Percentile	95th Percentile
toilet flushing	10	$0.0276 \times P + 61.782$	$-7 \times 10^{-6} \times P^2 + 0.0379 \times P + 56.864$	$8 \times 10^{-6} \times P^2 + 0.0148 \times P + 68.685$
	15	$0.0299 \times P + 64.589$	$-8 \times 10^{-6} \times P^2 + 0.0445 \times P + 57.073$	$8 \times 10^{-6} \times P^2 + 0.0191 \times P + 69.642$
	20	$0.0316 \times P + 66.804$	$-1 \times 10^{-5} \times P^2 + 0.0505 \times P + 57.233$	$1 \times 10^{-5} \times P^2 + 0.0164 \times P + 74.115$
garden irrigation	10	$0.0183 \times P + 41.614$	$-6 \times 10^{-6} \times P^2 + 0.0271 \times P + 36.728$	$6 \times 10^{-6} \times P^2 + 0.0104 \times P + 46.063$
	15	$0.0200 \times P + 51.705$	$-9 \times 10^{-6} \times P^2 + 0.0352 \times P + 43.927$	$8 \times 10^{-6} \times P^2 + 0.0087 \times P + 57.493$
	20	$0.0214 \times P + 61.223$	$-8 \times 10^{-6} \times P^2 + 0.0338 \times P + 54.87$	$7 \times 10^{-6} \times P^2 + 0.0113 \times P + 66.569$
toilet flushing and garden irrigation	10	$0.0233 \times P + 38.775$	$-7 \times 10^{-6} \times P^2 + 0.0335 \times P + 33.891$	$8 \times 10^{-6} \times P^2 + 0.0125 \times P + 44.048$
	15	$0.0282 \times P + 38.482$	$-9 \times 10^{-6} \times P^2 + 0.0424 \times P + 31.477$	$9 \times 10^{-6} \times P^2 + 0.0153 \times P + 44.332$
	20	$0.0320 \times P + 38.508$	$-9 \times 10^{-6} \times P^2 + 0.0466 \times P + 31.437$	$9 \times 10^{-6} \times P^2 + 0.0185 \times P + 45.036$

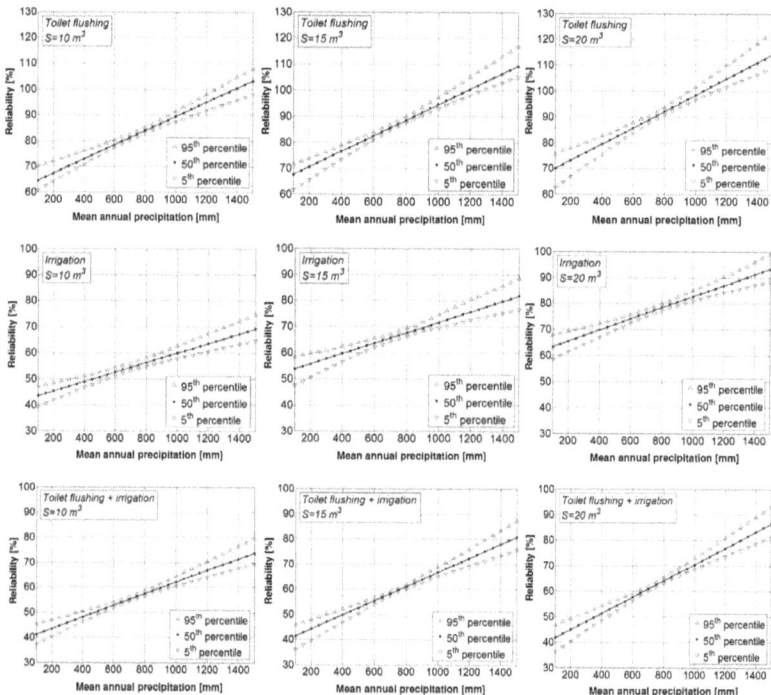

Figure 7. Reliability curves and their uncertainty bands for each uses and analyzed storage volumes.

In every case, the width of the uncertainty bands increases for the lowest and highest values of mean annual precipitation. In the case of the lowest values, the uncertainty is related to the fact that the reliability of the system is considerably affected by the amount of harvested rainwater, because of the potential failure of the RWH system in meeting the water demand. In the case of the highest values, the uncertainty in the reliability is related to the fact that the amount of harvested water is likely to exceed the water demand. The installation of an RWH system in the above mentioned cases requires a deeper analysis to verify its cost-effectiveness. Where the amount of rainwater is not enough to meet the water demand, the analyzed volumes and collection surfaces are not adequate to ensure a high level of water savings, making households dependent on other water sources for most or part of the year. On the other hand, where the amount of rainwater exceeds the needs of the household, the rainwater that overflows the storage tank represents an economic loss because this water could meet other demands, allowing a greater independence from the traditional supply system and, therefore, further savings.

3.5. Cost-Benefit Analysis

An economic analysis of the RWH system was carried out in order to investigate the balance between the investment/cost for system purchase and installation, and the benefits obtained by the rainwater use for the three considered demands. To this aim, a schematic underground installation of an RWH system was considered, consisting of a pre-fabricated concrete tank provided with a first flush device, a manhole with a rainwater filter, a pumping system and its Programmable Logic Controller (PLC) equipment, the drainage piping system inlet and outlet, the tank, and the piping distribution system to supply the rainwater for the analyzed uses (Figure 8). Table 4 summarizes the costs of the RWH system elements for each tank capacity and each use. These costs have been obtained starting from the unit rates, drawn from the official regional price list for civil infrastructures [48], and by means of a market survey.

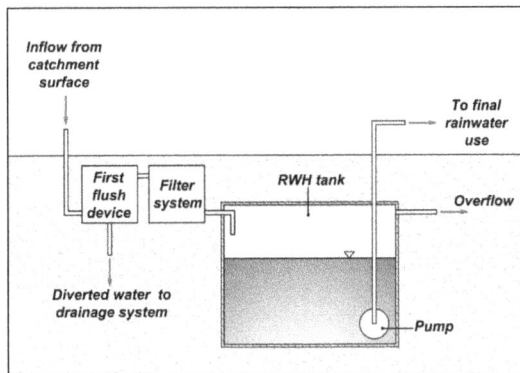

Figure 8. Schematic underground installation of the RWH system.

The tank purchase and installation highly affects the total RWH system cost (Table 4), as confirmed by different studies in the literature [49–51]. Moreover, the RWH system for toilet flushing is more expensive than that for only irrigation use, due to the installation costs related to the piping distribution system in the building.

In the present analysis, the costs related to the system maintenance were considered negligible when compared to purchase and installation costs [50]. With regard to operation costs and, in particular, the energy costs needed to pump the rainwater for the analyzed uses, these costs were neglected. Regarding this assumption, some considerations have to be made. In most of the sites in Sicily, water managers often adopt the intermittent distribution to cope with water shortage periods or to contain high water losses, due to the lack of adequate maintenance of the supply networks [52,53]. As a consequence, the plumbing systems of households are frequently equipped with pumping stations and private tanks to collect potable water during service periods and supply water when the service is not available. Because of the lack of confidence of users on the reliability of the water supply service, the private tanks and the pumping system are not bypassed, even if the distribution system operates on a continuous basis. Thus, the users are prepared for unexpected interruption of the supply service.

Therefore, in most of the sites of Sicily, users nowadays have to pay a large amount for energy needed to draw water from the public network because of private storage tanks and pumping systems [54].

Table 4. Elements costs of a schematic RWH system for each tank size and rainwater use.

Item	Toilet Flushing			Irrigation			Toilet Flushing + Irrigation		
Tank capacity [m³]	10	15	20	10	15	20	10	15	20
Cost for concrete tank purchase, the first flush device and their underground placing	€ 1778	€ 2284	€ 2991	€ 1700	€ 2284	€ 2991	€ 1700	€ 2284	€ 2991
Pipes drainage system inlet and outlet tank	€ 178	€ 178	€ 178	€ 178	€ 178	€ 178	€ 178	€ 178	€ 178
Piping system for not potable water supply	€ 194	€ 194	€ 194	€ 290	€ 290	€ 97	€ 290	€ 290	€ 290
Pump and PLC equipment	€ 2000	€ 2000	€ 2000	€ 2000	€ 2000	€ 1500	€ 2000	€ 2000	€ 2000
Rainwater filter	€ 220	€ 220	€ 250	€ 220	€ 220	€ 250	€ 220	€ 220	€ 250
Total costs	€ 4370	€ 4876	€ 5612	€ 4388	€ 4973	€ 5016	€ 4388	€ 4973	€ 5709

With regard to the benefits related to the RWH system installation, only the benefits due to the potable water saving have been considered. In particular, the financial benefit has been evaluated in terms of reduction of the annual water bill from water utilities. Even if relevant, in this analysis the environmental and social benefits have not been accounted. The cost-benefit analysis has been carried out according to the *"Guide to cost-benefit analysis of investment projects"* in Europe [55]. Namely, two performance indicators, the Net Present Value (*NPV*) and the payback period (*PBP*), have been evaluated, as described by Khastagir and Jayasuriya [50] and Matos *et al.* [56]. In the analysis, some assumptions have been made:

- The evaluation period to assess the *NPV* has been set equal to 20 years [8,56,57];
- according to [55], a discount rate of 5% has been assumed;
- the inflation rate has been assumed equal to 8% (on the basis of the inflation rate of potable water price in Italy in recent years);
- the actual price for potable water has been set equal to 2.5 €/m³ (obtained as the average of the actual prices adopted by different water utilities operating in Sicily).

The effect of the variability of annual yield related to the different location of the system installation has been accounted for in the *PBP* and the *NPV* appraisal, considering the minimum, the maximum and the mean annual yield in the area of study. Results are shown in Table 5 for each tank size and rainwater use. As expected, for a given use, the payback period increases with the tank capacity. For the toilet flushing use, a 10 m³ tank capacity was found to to be adequate, since an increase of the tank size of 5–10 m³ improves the system R_V of only the 1%. For a yield equal to the mean annual value, the payback period is 21 years (closer to the assumed evaluation period). As regards to irrigation use, the annual benefits are scarce, due to the lower annual yield values. As a consequence, payback periods are higher than the assumed evaluation period, specifically about 34 years for the three annual yield values, meaning that 20 years are enough to get back only half of the costs of system installation. In terms of annual R_V, a 20 m³ capacity was found to be a feasible solution for this use. For both uses, the payback period related to the mean annual yield are similar, as well as the system R_V. Therefore, in this case, the 10 m³ capacity seems to be the most advantageous.

Table 5. *NPV* and *PBP* values related to each tank size and different annual yields for each rainwater use.

Rainwater Use	Tank Volume	Investments/ Costs	Annual Yield/ Water Saving		Annual R_V	NPV (20 Years)	$PBP = N_{CER}$
	(m³)	(€)	(m³/year)		(%)	(€)	(year)
Toilet flushing	10	€ 4388	*max*	78	100%	€ 1,137	17
			mean	60	77%	−€ 134	21
			min	34	43%	−€ 1969	31
	15	€ 4973	*max*	78	100%	€ 631	19
			mean	61	78%	−€ 569	22
			min	34	43%	−€ 2476	34
	20	€ 5709	*max*	78	100%	−€ 106	21
			mean	61	78%	−€ 1306	25
			min	34	43%	−€ 3212	37
Irrigation	10	€ 3773	*max*	38	86%	−€ 1090	26
			mean	25.6	58%	−€ 1965	34
			min	13.4	30%	−€ 2827	50
	15	€ 4279	*max*	44.4	100%	−€ 1596	29
			mean	30.7	69%	−€ 2112	33
			min	15.2	34%	−€ 3206	50
	20	€ 5016	*max*	44.4	100%	−€ 1881	29
			mean	34.7	78%	−€ 2566	34
			min	15.2	34%	−€ 3943	55
Toilet flushing + Irrigation	10	€ 4370	*max*	94.5	77%	€ 2283	15
			mean	63.7	52%	€ 109	20
			min	33.9	28%	−€ 1995	32
	15	€ 4876	*max*	104.5	85%	€ 1699	16
			mean	65.5	53%	−€ 348	22
			min	33.9	28%	−€ 2579	34
	20	€ 5612	*max*	109.5	89%	€ 2022	16
			mean	66.5	54%	−€ 1014	24
			min	33.9	28%	−€ 3316	38

4. Conclusions

For a long time, urban design and planning has ignored the advantages of RWH as a sustainable water resources management tool; however, interest in RWH systems as an alternative water source has recently increased. These systems can provide a supplementary water supply in urbanized areas when integrated with existing conventional water supply systems, or they can serve as the main water source in rural areas where the availability of water resources is a critical issue. Moreover, utilizing RWH represents an effective adaptive strategy to climate change against the reduction of water availability. The feasibility of rainwater harvesting in a particular locality is highly dependent on rainfall characteristics (intensity and frequency). Other variables, such as catchment area and type of catchment surface, usually can be modified to improve system performance.

In this study, a behavioral model was applied to assess the performance of an RWH system in terms of its reliability. Water demand for toilet flushing and garden irrigation and three years of historical daily rainfall data for 111 locations in Sicily were used as input to the system simulation model, the YAS algorithm. The analysis of simulation results, in terms of annual reliability of the RWH system, highlighted the possibility of obtaining good performances when the collected water is intended solely for toilet flushing. In this case, the saving of water from other supply systems makes the RWH system to be cost-effective in most of the analyzed territory. In particular, a storage capacity of 20 m³ is able to ensure the complete meeting of water demand for toilet flushing in a wide northern area of Sicily. On the other hand, the use of rainwater for garden irrigation requires, in most of the island, higher storage capacities in order to obtain advantageous performances in terms of water saving. Due to the different temporal patterns of water demands, the coupling of the two uses, toilet flushing and garden irrigation, is not particularly advantageous for the considered storage volumes and collection surfaces.

The analysis of the monthly variability of the RWH system's reliability showed that the temporal variability of rainfall over the year has an important impact on storage volume. In an area with uniform monthly precipitation throughout the year, a smaller storage volume is necessary than that required in an area with a distinct seasonal precipitation distribution.

Results from the application of the YAS algorithm to different sites in Sicily were used to analyze the correlation between mean annual precipitation and the reliability of the examined RWH system. The analysis defined curves that are valid for the entire area of study and relate to the above mentioned variables. The equations of these curves represent a useful tool for practical application in Sicily, easily and quickly providing a value of the RWH reliability corresponding to a given value of mean annual rainfall. The uncertainty related to the obtained curves was assessed by reducing the original dataset and obtaining alternative curves. Future research can assess the implications of household occupancy and the impacts of rooftop and courtyard areas and storage capacity on reliability. These factors can then be integrated into the proposed equations to obtain general relationships to more effectively evaluate the performance of any RWH system.

A cost-benefit analysis has been performed, providing the Net Present Value and the payback periods on the capital cost of system installation. Results enabled the identification of the most feasible tank capacity. Despite the high payback periods of capital cost, the environmental and social advantages related to the use of RWH systems cannot be neglected. Indeed, these systems promote a more sustainable water use and a greater resilience to water scarcity.

Further analysis should also account for the effect of climate change on precipitation. The equations presented here are valid under the assumption that the mean annual precipitation will not be affected by variations in the next years. The existence of trends could significantly affect the performance of an RWH system. Specifically, the reduction of rainfall amount and the variation of rainfall temporal distribution over the year (in particular the concentration of annual rainfall in short periods) could lead to a considerable decrease of the system efficiency. Therefore, the design of RWH tanks should also involve an analysis of future climate scenarios derived from regional climate models.

In summary, RWH systems can play an important role in supplementing conventional water supply systems. For this reasons, incentives and government support could be important to encourage householders to adopt RWH water systems in residential urban areas.

Author Contributions: Author Contributions: All the authors contributed equally to this work.

Conflicts of Interest: Conflicts of Interest: The authors declare no conflict of interest.

References

1. Bates, B.; Kundzewicz, Z.; Wu, S.; Palutikof, J. Observed and projected changes in climate as they relate to water. *Climate Change and Water*; Intergovernmental Panel on Climate Change: Geneva, Switzerland, 2008; pp. 13–31.
2. Tsiourtis, N.X. Desalination and the environment. *Desalination* **2001**, *141*, 223–236. [CrossRef]
3. Liaw, C.H.; Tsai, Y.L. Optimum storage volume of rooftop rain water harvesting systems for domestic use. *J. Am. Water Resour. Assoc.* **2004**, *40*, 901–912. [CrossRef]
4. Gould, J.; Niessen-Peterson, E. *Rainwater Catchment Systems for Domestic Supply: Design, Construction and Implementation*; Intermediate Technology: London, UK, 1999.
5. Pandey, D.N.; Gupta, A.K.; Anderson, D.M. Rainwater harvesting as an adaptation to climate change. *Curr. Sci.* **2003**, *85*, 46–59.
6. AbdelKhaleq, R.; Ahmed, A. Rainwater harvesting in ancient civilizations in Jordan. *Water Sci. Technol. Water Suppl.* **2007**, *7*, 85–93. [CrossRef]
7. Kahinda, J.M.M.; Taigbenu, A.E.; Boroto, J.R. Domestic rainwater harvesting to improve water supply in rural South Africa. *Phys. Chem. Earth* **2007**, *32*, 1050–1057. [CrossRef]
8. Farreny, R.; Gabarrell, X.; Rieradevall, J. Cost-efficiency of rainwater harvesting strategies in dense Mediterranean neighbourhoods. *Resour. Conserv. Recycl.* **2011**, *55*, 686–694. [CrossRef]
9. Morales-Pinzón, T.; Lurueña, R.; Gabarrell, X.; Gasol, C.M.; Rieradevall, J. Financial and environmental modelling of water hardness—Implications for utilizing harvested rainwater in washing machines. *Sci. Total Environ.* **2014**, *470*, 1257–1271. [CrossRef] [PubMed]

10. Förster, J. Variability of roof runoff quality. *Water Sci. Technol.* **1999**, *39*, 137–144. [CrossRef]
11. Göbel, P.; Dierkes, C.; Coldewey, W.G. Storm water runoff concentration matrix for urban areas. *J. Contam. Hydrol.* **2007**, *91*, 26–42. [CrossRef] [PubMed]
12. Adeniyi, I.F.; Olabanji, I.O. The physico-chemical and bacteriological quality of rainwater collected over different roofing materials in Ile-Ife, southwestern Nigeria. *Chem. Ecol.* **2005**, *21*, 149–166. [CrossRef]
13. Melidis, P.; Akratos, C.S.; Tsihrintzis, V.A.; Trikilidou, E. Characterization of rain and roof drainage water quality in Xanthi, Greece. *Environ. Monit. Assess.* **2007**, *127*, 15–27. [CrossRef] [PubMed]
14. Chang, M.; McBroom, M.W.; Beasley, R.S. Roofing as a source of nonpoint water pollution. *J. Environ. Manag.* **2004**, *73*, 307–315. [CrossRef] [PubMed]
15. Van der Sterren, M.; Rahman, A.; Dennis, G.R. Quality and quantity monitoring of five rainwater tanks in Western Sydney, Australia. *J. Environ. Eng.* **2013**, *139*, 332–340. [CrossRef]
16. Jones, M.; Hunt, V.F.; Wright, J. Rainwater harvesting experiences in the humid south-east USA. In Proceedings of the World Environment and Water Resources Congress, Kansas City, MO, USA, 17–21 May 2009.
17. Jones, M.P.; Hunt, W.F. Performance of rainwater harvesting systems in the southeastern United States. *Resour. Conserv. Recycl.* **2010**, *54*, 623–629. [CrossRef]
18. Zhang, Y.; Grant, A.; Sharma, A.; Chen, D.; Chen, L. Alternative water resources for rural residential development in Western Australia. *Water Resour. Manag.* **2010**, *24*, 25–36. [CrossRef]
19. Villarreal, E.L.; Dixon, A. Analysis of a rainwater collection system for domestic water supply in Ringdansen, Norrköping, Sweden. *Build. Environ.* **2005**, *40*, 1174–1184. [CrossRef]
20. Ghisi, E.; Bressan, D.L.; Martini, M. Rainwater tank capacity and potential for potable water savings by using rainwater in the residential sector of southeastern Brazil. *Build. Environ.* **2007**, *42*, 1654–1666. [CrossRef]
21. Panigrahi, B.; Panda, S.N.; Mal, B.C. Rainwater conservation and recycling by optimal size on-farm reservoir. *Resour. Conserv. Recycl.* **2007**, *50*, 459–474. [CrossRef]
22. Guo, Y.; Baetz, B.W. Sizing of rainwater storage units for green building applications. *J. Hydrol. Eng.* **2007**, *12*, 197–205. [CrossRef]
23. Ahiablame, L.; Engel, B.; Venort, T. Improving water supply systems for domestic uses in Urban Togo: The case of a suburb in Lomé. *Water* **2012**, *4*, 123–134. [CrossRef]
24. Woltersdorf, L.; Liehr, S.; Döll, P. Rainwater harvesting for small-holder horticulture in Namibia: Design of garden variants and assessment of climate change impacts and adaptation. *Water* **2015**, *7*, 1402–1421. [CrossRef]
25. Abdulla, F.A.; Al-Shareef, A.W. Roof rainwater harvesting systems for household water supply in Jordan. *Desalination* **2009**, *243*, 195–207. [CrossRef]
26. Kadam, A.K.; Kale, S.S.; Pande, N.N.; Pawar, N.J.; Sankhua, R.N. Identifying potential rainwater harvesting sites of a semi-arid, basaltic region of Western India, using SCS-CN method. *Water Resour. Manag.* **2012**, *26*, 2537–2554. [CrossRef]
27. Liaw, C.H.; Chiang, Y.C. Dimensionless Analysis for Designing Domestic Rainwater Harvesting Systems at the Regional Level in Northern Taiwan. *Water* **2014**, *6*, 3913–3933. [CrossRef]
28. Hanson, L.S.; Vogel, R.M.; Kirshen, P.; Shanahan, P.; Starrett, S. Generalized Storage-Reliability-Yield Equations for Rainwater Harvesting Systems. In Proceedings of the world environmental and water Resources Congress, Kansas City, MO, USA, 17–21 May 2009.
29. Hajani, E.; Rahman, A. Reliability and cost analysis of a rainwater harvesting system in peri-urban regions of Greater Sydney, Australia. *Water* **2014**, *6*, 945–960. [CrossRef]
30. Sazakli, E.; Alexopoulos, A.; Leotsinidis, M. Rainwater harvesting, quality assessment and utilization in Kefalonia Island, Greece. *Water Res.* **2007**, *41*, 2039–2047. [CrossRef] [PubMed]
31. Campisano, A.; Modica, C. Optimal sizing of storage tanks for domestic rainwater harvesting in Sicily. *Resour. Conserv. Recycl.* **2012**, *63*, 9–16. [CrossRef]
32. Wisner, P.; P'ng, J.C. OTTHYMO, A Model for Master Drainage Plans, IMPSWM Urban Drainage Modelling Procedures, 2nd ed.; Department of Civil Engineering, University of Ottawa: Ottawa, ON, Canada, 1983.
33. Khastagir, A.; Jayasuriya, N. Optimal sizing of rain water tanks for domestic water conservation. *J. Hydrol.* **2010**, *381*, 181–188. [CrossRef]
34. Thornton, R.C.; Saul, A.J. Some quality characteristics of combined sewer flows. *J. Public Health Eng.* **1986**, *14*, 35–38.

35. Yaziz, M.I.; Gunting, H.; Sapari, N.; Ghazali, A.W. Variations in rainwater quality from roof catchments. *Water Res.* **1989**, *23*, 761–765. [CrossRef]

36. Coombes, P. Rainwater Tanks Revisited: New Opportunities for Urban Water Cycle Management. Ph.D. Thesis, University of Newcastle, NSW, Australia, 2002.

37. Lizárraga-Mendiola, L.; Vázquez-Rodríguez, G.; Blanco-Piñón, A.; Rangel-Martínez, Y.; González-Sandoval, M. Estimating the Rainwater Potential per Household in an Urban Area: Case Study in Central Mexico. *Water* **2015**, *7*, 4622–4637. [CrossRef]

38. Thornthwaite, C.W. An approach toward a rational classification of climate. *Geog. Rev.* **1948**, 55–94. [CrossRef]

39. Allen, R.G.; Pereira, L.S.; Raes, D.; Smith, M. Crop evapotranspiration: Guidelines for computing crop water requirements. In *Irrigation and Drainage Paper*; FAO: Rome, Italy, 1998; Volume 56, p. 300.

40. Fry, J.D.; Huang, B. *Applied Turfgrass Science and Physiology*; John Wiley & Sons: Hoboken, NJ, USA, 2004.

41. Fu, J.; Dernoeden, P.H. Creeping Bentgrass Putting Green Turf Responses to Two Irrigation Practices: Quality, Chlorophyll, Canopy Temperature, and Thatch-Mat. *Crop Sci.* **2009**, *49*, 1071–1078. [CrossRef]

42. Jordan, J.E.; White, R.H.; Vietor, D.M.; Hale, T.C.; Thomas, J.C.; Engelke, M.C. Effect of irrigation frequency on turf quality, shoot density, and root length density of five bentgrass cultivars. *Crop Sci.* **2003**, *43*, 282–287. [CrossRef]

43. Irrigation Association and American Society of Irrigation Consultants. *Landscape Irrigation Best Management Practices*; Irrigation Association and American Society of Irrigation Consultants: Fairfax, VA, USA, 2014.

44. Marchione, V. Effetti della riduzione dei volumi irrigui su specie microterme da tappeto erboso in un ambiente della Puglia. *Ital. J. Agron. River Agron.* **2009**, *4*, 961–966.

45. Fewkes, A.; Butler, D. Simulating the performance of rainwater collection and reuse systems using behavioural models. *Build. Serv. Eng. Res. Technol.* **2000**, *21*, 99–106. [CrossRef]

46. Ward, S.; Memon, F.A.; Butler, D. Rainwater harvesting: Model-based design evaluation. *Water Sci. Technol.* **2010**, *61*, 85–96. [CrossRef] [PubMed]

47. Jenkins, D.; Pearson, F.; Moore, E.; Sun, J.K.; Valentine, R. *Feasibility of Rainwater Collection Systems in California*; Contribution No. 173; Californian Water Resources Centre, University of California: Davis, CA, USA, 1978.

48. Gazzetta Ufficiale Regione Sicilia. *Prezzario Unico Regionale per i Lavori Pubblici 2013 Della Regione Sicilia*; Gazzetta Ufficiale Regione Sicilia: Palermo, Italy, 2013. (In Italian)

49. Chilton, J.C.; Maidment, G.G.; Marriott, D.; Francis, A.; Tobias, G. Case Study of a rainwater harvesting system in a commercial building with a large roof. *Urban Water* **2000**, *1*, 345–354. [CrossRef]

50. Khastagir, A.; Jayasuriya, N. Investment evaluation of rainwater tanks. *Water Resour. Manag.* **2011**, *25*, 3769–3784. [CrossRef]

51. Ghisi, E.; Schondermark, P.N. Investment feasibility analysis of rainwater use in residence. *Water Resour. Manag.* **2013**, *27*, 2555–2576. [CrossRef]

52. De Marchis, M.; Fontanazza, C.M.; Freni, G.; la Loggia, G.; Napoli, E.; Notaro, V. A model of the filling process of an intermittent distribution network. *Urban Water* **2010**, *7*, 321–333. [CrossRef]

53. De Marchis, M.; Fontanazza, C.M.; Freni, G.; la Loggia, G.; Napoli, E.; Notaro, V. Analysis of the impact of intermittent distribution by modelling the network-filling process. *J. Hydroinform.* **2011**, *13*, 358–373. [CrossRef]

54. Fontanazza, C.M.; Freni, G.; la Loggia, G.; Notaro, V.; Puleo, V. Evaluation of the water scarcity energy cost for users. *Energies* **2013**, *6*, 220–234. [CrossRef]

55. European Commission, Directorate General Regional Policy. Guide to Cost-Benefit Analysis of Investment projects—Structural Funds, Cohesion Fund and Instrument for Pre-Accession. Available online: http://ec.europa.eu/regional_policy/sources/docgener/guides/cost/guide2008_en.pdf (accessed on 12 November 2015).

56. Matos, C.; Bentes, I.; Santos, C.; Imteaz, M.; Pereira, S. Economic analysis of a rainwater harvesting system in a commercial building. *Water Resour. Manag.* **2015**, *29*, 3971–3986. [CrossRef]

57. Zhang, Y.; Chen, D.; Chen, L.; Ashbolt, S. Potencial for rainwater use in high-rise buildings in Australian cities. *J. Environ. Manag.* **2009**, *91*, 222–226. [CrossRef] [PubMed]

water

MDPI

Article

Rainwater Harvesting Typologies for UK Houses: A Multi Criteria Analysis of System Configurations

Peter Melville-Shreeve *, Sarah Ward and David Butler

Centre for Water Systems, University of Exeter, Exeter EX4 4QF, UK; sarah.ward@exeter.ac.uk (S.W.); d.butler@exeter.ac.uk (D.B.)
* Correspondence: pm277@exeter.ac.uk; Tel.: +44-1392-723600

Academic Editor: Ataur Rahman
Received: 6 October 2015; Accepted: 25 March 2016; Published: 1 April 2016

Abstract: Academic research and technological innovation associated with rainwater harvesting (RWH) systems in the UK has seen a shift of emphasis in recent years. Traditional design approaches use whole life cost assessments that prioritise financial savings associated with the provision of an alternative water supply. However, researchers and practitioners are increasingly recognising broader benefits associated with rainwater reuse, such as stormwater attenuation benefits. This paper identifies and describes a set of novel RWH system configurations that have potential for deployment in UK houses. Conceptual schematics are provided to define these innovations alongside traditional configurations. Discussion of the drivers supporting these configurations illustrates the opportunities for RWH deployment in a wide range of settings. A quantitative multi criteria analysis was used to evaluate and score the configurations under a range of emerging criteria. The work identifies several RWH system configurations that can outperform traditional ones in terms of specified cost and benefits. Selection of a specific RWH technology is shown to be highly dependent on user priorities. It is proposed that the system configurations highlighted could enable RWH to be cost-effectively installed in a broad set of contexts that have experienced minimal exploitation to date.

Keywords: configurations; decision support; multi criteria analysis; product innovation; rainwater harvesting; sustainable drainage systems; source control

---

1. Introduction

1.1. Rainwater Harvesting at UK Houses

Rainwater harvesting (RWH) systems in the UK have traditionally been installed at domestic residences for the single objective of providing a non-potable water supply for use in toilets, laundry facilities and for garden irrigation [1,2]. Unlike some fully off-grid configurations implemented elsewhere [3,4], system configurations in the UK are supplemented by mains water supplies for potable water applications such as drinking, bathing and dishwashing. Germany has seen strong uptake of RWH technologies as reported by Partzsch [5] with 80,000 installations per annum and a total industry value of 340 million Euros. With successful growth in that market driven by policies that seek to (financially) support green technologies, one in three houses constructed in 2005 installed a rainwater tank. However, the nascent UK RWH installation market has developed with early-adopters purchasing well-established technologies that directly derive from installations found in countries where RWH is now mainstream, such as Germany [6] and Australia [7].

In fact, a review of three leading RWH system providers in the UK illustrates that they either license products from European manufacturers or have mimicked such configurations [8–10]. Whilst suitable for some sites, the direct transplantation of these off-the-shelf, traditional RWH system configurations into the UK marketplace could prevent optimal RWH solutions from being installed, as the current market-place only offers a limited range of technologies to potential purchasers. Additionally, these traditional RWH systems are best suited to new build houses with large gardens or driveways (under which tanks can be placed) with high non-potable water consumption. They can be difficult and costly to retrofit and may have high maintenance requirements [11]. House building trends in the UK are for smaller properties with low-flush toilets and less garden space. Recent research on water using practices revealed that 62% of the sample had some garden applications for which rainwater could be used (plants, flowers, lawn). However, 26% of this subset did not irrigate or water their gardens, but simply waited for rain [12]. In combination, this means that there is a growing need for retrofitable RWH systems, which utilise smaller rainwater tanks. However, there are few commercially-available systems to address this opportunity. Furthermore, optimal RWH systems might be designed to respond to a wider set of drivers than simply achieving (non-potable) water supply, such as reducing total water related energy consumption and improving stormwater control.

Minimal government incentives, subsidy or support for RWH means the UK market remains nascent. At the residential property scale, installation rates remain low with the market reportedly worth just £8 million in 2009 [13]. This is no doubt due to the whole life cost benefits of traditional technologies resulting in long payback periods to individual purchasers [14]. There is therefore a compelling case to develop an affordable, retrofittable and multi-benefit range of RWH system configurations and options to respond to these property and regime level drivers.

In this paper, traditional and innovative RWH systems have been identified and their configurations described. Secondly, a set of criteria are defined that enable RWH system configurations to be evaluated using multi criteria analysis (MCA). The outputs from the research illustrate the ability of RWH systems to achieve a number of objectives and the methods are intended to support designers, householders, water companies and installers in understanding the broader opportunities presented by emerging innovative RWH technologies.

1.2. Existing Cost–Benefit Approaches to RWH Assessment

A straightforward method of financial appraisal can be achieved by evaluating the payback period for a RWH system. This sets the capital cost against the long-term savings generated from the reduced water supply and associated sewerage costs. Contemporary RWH studies and modelling tools also integrate the operational costs and planned maintenance costs (for example pump replacement and tank cleaning) [15]. Such an approach was demonstrated by Roebuck *et al.* [14], who concluded that a whole life cost (WLC) approach is most appropriate for undertaking financial appraisal of RWH systems in the UK. This work advocated the need to include capital, maintenance, operational and decommissioning costs while attributing financial benefits to the savings linked to water and sewerage tariff reductions. Ward *et al.* [11] agree that WLC approaches represent best practice and propose that daily rainfall datasets should be deployed to enable more accurate modelling of RWH systems [16]. Roebuck and colleagues' later work [17] also illustrated that use of simplified tools (for example those that do not account for WLC) can result in designs that have hypothetically viable payback periods but cost more to maintain and operate than they save when whole lifecycle costs are included.

A wider review of literature and RWH system design tools illustrates that appraisal beyond financial benefit is lacking [18–20]. An appraisal under a single objective "maximise whole-life financial benefit of water reuse" omits many of the nuanced benefits offered by RWH systems. Consequently, examination of novel RWH system configurations benchmarked against a wider set of criteria is warranted.

1.3. A Framework for RWH Evaluation under a Range of Criteria

Following Coombes [21], the work set out in this paper develops a decision space that trades off whole life benefits and whole life costs. This concept neatly frames the need for innovation in the context of the UK's RWH industry through visualizing system configurations using a Pareto front. The delivery of optimal water management is currently constrained by the size and variety of the original set of solutions at the designer's disposal. For example, if the designer of a new housing development seeks to install a water reuse system, they might reasonably investigate the relevant British Standards; BS8595:2013 Code of practice for the selection of water reuse systems [22] and BS8515:2009+A1:2013 Rainwater Harvesting Systems—Code of practice [2]. The components and configurations included within the standards might be extracted and evaluated on a case-by-case basis using a handful of cost benefit metrics. These designs represent the total set of potential design solutions. The designer may conclude that RWH is not a cost effective option, as no solutions evaluated met the designer's budgetary constraints. Consequently, the initial target to incorporate water reuse into the development remains unmet. In graphical form, this is conceptualised in Figure 1. It is evident from this graphic that expanding the original set of solutions can increase the likelihood that suitable RWH system configurations can be identified. In this example, Figure 1 identifies that two previously "unseen" solutions are available to the designer that are within budget but were not considered in the previously limited decision space. It is proposed that the development of a quantitative RWH assessment tool that incorporates a range of criteria will enable practitioners to widen the decision space and implement RWH systems in locations where single objective benefit appraisals fail to satisfy cost benefit criteria.

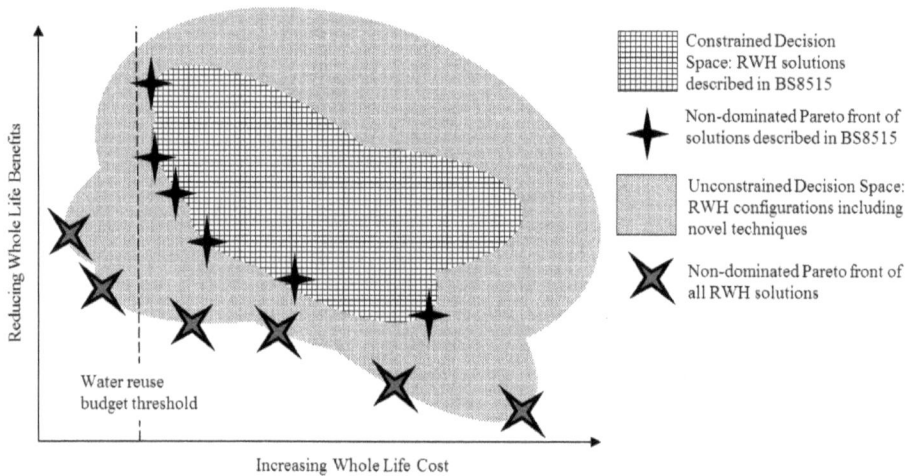

Figure 1. Conceptualising the benefits of innovative design of novel RWH system configurations (adapted from Coombes [21]).

1.4. Multi Criteria Analysis

Multi criteria analysis (MCA) methods are frequently used in the field of integrated water management to support decision makers who wish to differentiate between options with complex, multi-faceted characteristics [23–26]. The methods typically follow a structure as follows [27]: define the problem, identify alternative options, define criteria and associated objectives, populate performance matrix and evaluate performance against criteria. With values for each criteria defined, previous studies have deployed a range of methods such as weighted summation or range of value methods [25]. These techniques add a final tier of expert judgement to enable the preferred option to be selected from those which perform strongly. The MCA method used in this paper describes innovative systems and identifies a process for differentiating between them. In a final step, three scenarios are defined to assess technology selection based on user preferences.

2. Method

The method adopted in this paper is a simple linear weighted MCA based on the following 6 steps, adapted from [27].

Step 1—Define the problem and associated parameters. A well-defined problem statement was needed to enable the MCA to be developed.

Step 2—Identify alternative options. A comprehensive literature review of existing and emerging RWH system configurations was conducted to identify and define their characteristics.

Step 3—Define criteria and associated objectives. The literature review identified a number of drivers (objectives) which have enabled five criteria to be defined for RWH implementation at a household level in the UK. Details of technologies and criteria were established from a broad range of sources, which included: patent searches, meetings with industry suppliers, site visits, conference attendance, facilitating workshops, innovation events with rainwater practitioners, collaborative design partnerships and reviews of industry texts and the peer-reviewed literature.

Step 4—Populate performance matrix. With system configurations and criteria defined, quantitative methods were used to evaluate each configuration.

Step 5—Evaluate performance against criteria. Results were generated to benchmark the configurations against one another to demonstrate how they perform when each criteria is considered as the sole objective in selecting a RWH configuration.

Step 6—Scenario Testing. Three hypothetical scenarios were defined and weightings allocated to the MCA in order to illustrate the effectiveness of the MCA approach as a decision support tool.

The UK has seen many developments and innovations in RWH design configurations, both within the RWH industry and within the academic research community [28–32]. The identification of details of these systems form the basis of Step 2. A summary of traditional and innovative RWH configurations is described in Section 3.2. A matrix was constructed to allow values for each configuration to be derived from literature or calculated against the five criteria determined in Section 3.3. The criteria were utilised to evaluate the configurations against each objective. To achieve this, a fixed set of parameters was used to define a case study house against which each RWH system could be assessed using a time-series model. For simplicity, the paper illustrates how the systems compare when assessed against a single house. The intention is that the method can be further utilised in order to allow decision makers to assess the range of RWH systems against any site.

3. RWH System Configurations and Drivers

This section describes the process of applying the previously described 6 MCA steps using the RWH industry as its focus.

3.1. Step 1—Define the Problem and Associated Parameters

The method set out in this paper seeks to answer the following problem statement:

"Identify a quantitative method to evaluate the broad benefits of a range of traditional and novel RWH technologies at a given location."

A set of fixed parameters was generated to enable comparison of RWH technologies to be undertaken at a domestic property. Parameters for a typical UK house are described in Table 1. The property is assumed to have: a pitched roof with a plan area of 60 m^2, four occupants utilising 150 L/person/day (with a usage ratio based on existing literature [33]), space and structural capacity for up to 2 No. 0.25 m^3 loft or wall mounted header tanks, and can accommodate up to 5 m^3 of above ground or below ground storage.

Table 1. Defining the characteristics for a typical UK house.

Model Parameter	Reference	Value
Roof Area (m^2)	User Selected	60
Roof Runoff Coefficient	User Selected	0.9
First-Flush Losses (L/day)	User Selected	5
Usage Ratio (WC:Laundry:Potable:Other)	[33]	30:20:5:45
Tank storage size	User Selected	• 0.5 m^3 if located in loft • 0.5 m^3 if located externally for gravity feed • 5 m^3 if located at or below ground level. (Storage volume reduced to 4.5 m^3 where mains top up also enters storage tank)
Time-series rainfall data	Exeter, UK	Daily rainfall (mm) records for Exeter, UK

3.2. Step 2—Identifying Alternative Options: RWH System Configurations

RWH systems comprise a number of components, which typically include: gutter systems, filters, storage tanks, tank overflows, pumps, pressure vessels, pipework, valves, backup supply systems, sensors/float switches and electronic controllers. Details of these components can be identified through grey-literature available from RWH providers [8–10] and are described in BS8515:2009+A1:2013 [2]. Detailed descriptions of well-defined components are not included here as they are already suitably described in existing texts [34]. Existing literature describing RWH typologies chiefly focuses on a small number of potential configurations [6]. Furthermore, some terminology used does not match terms used by current UK RWH suppliers. The following typologies aim to extend and clarify these terminologies.

3.2.1. Best Practice in the UK: Traditional RWH System Configurations

In the UK, residential RWH systems typically utilise buried tanks although above ground tanks are also sometimes installed. Pumped flows are delivered via direct-feed or header tank systems. Consequently, four traditional RWH system configurations were identified as representing current best practice for household installations as described in Figure 2 [8–10]. The systems illustrated in Figure 2 each capture rainfall from the roof and store the filtered water in below ground (Figure 2a,c) or above ground (Figure 2b,d) tanks. Rainwater is then delivered by a submersible pump to non-potable applications either by direct-feed (Figure 2a,b) or via a header tank (Figure 2c,d). For the purposes of clarity, the overflow outlet from the system is described as a sewer (for example a combined sewer network) although RWH systems can also discharge to an infiltration device, surface water sewer or watercourse, depending on the site setting.

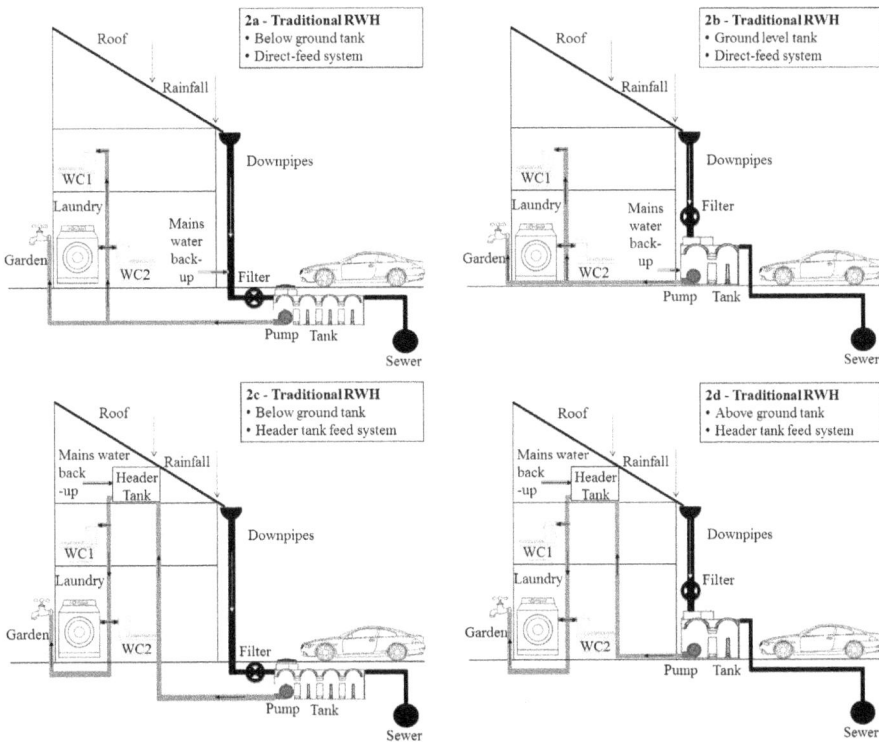

Figure 2. Conceptual schematics for four traditional rainwater harvesting RWH configurations used in the UK. (**a**) below ground, direct-feed ;(**b**) above ground, direct-feed ;(**c**)below ground, header tank feed ;(**d**) above ground, header tank feed.

3.2.2. Emerging Practice in UK: Innovative RWH System Configurations

In addition to the traditional RWH system configurations set out in Figure 2, a series of RWH innovations were identified. Through the collection of evidence, as described in Table 2, it is apparent that stormwater control potentially represents an additional key driver for innovation of RWH technologies. A summary of the innovations identified is set out in Table 2 and the configurations are diagrammatically illustrated in Figure 3.

Table 2. Innovative RWH system configurations.

System Provider and Patent No.	Description	Country	References
FlushRain Ltd., Farnham Common, UK. Patent: GB2449534	A patented suction pump system that captures rainwater from downpipes and stores rainwater in large header tanks. Easily retrofitted, with no external tanks.	UK	[35,36]
Aqua Harvest and Save Patent: GB2480834	A patented gutter-located pump system lifts rainwater into large header tanks. Easily retrofitted, with no external tanks.	UK	[36,37]
Atlas Water Harvesting Patent: GB2496729 and Rooftop Rain Patent:GB2475924 and GB2228521	A gravity-fed inlet is installed within the roof to enable ~50% of the roof to flow under gravity into large header tanks within the loft.	UK	[38–41]
Aqualogic (ARC): Rainbeetle GB2501313-B	An externally mounted tank, located near the roofline is installed to store rainwater and deliver flows by gravity.	UK	[42]
Hydromentum, Water Powered Technologies Ltd., Bude, UK.	An externally mounted header tank, drives a passively powered (zero electricity) pump to lift flows to a header tank.	UK	[43]
RainActiv, Rainwater Harvesting Ltd., Peterborough UK.	A passive rainwater discharge control (flow attenuation system) for inclusion within RWH tanks to ensure some storage is always maintained to attenuate extreme storm events.	Germany, USA, UK	[44–46]
KloudKeeper Ltd., Exeter, UK.	An active rainwater discharge control (flow attenuation system) for inclusion within RWH tanks to ensure some storage is always maintained to attenuate extreme storm events.	UK	[2]
IOTA,Melbourne South East Water (Aus) and Geosyntec, Boca Raton, FL, (USA)	A real-time control system that enables weather forecast data to support a decision maker to empty a RWH tank in a controlled way before a storm, thus ensuring capacity is available to capture extreme storm events.	Australia, Korea, USA	[47,48]
RainSafe, Newtown Mt Kennedy, Ireland	Rainwater treatment system that enables harvested rainwater to meet potable water standards.	UK	[49]

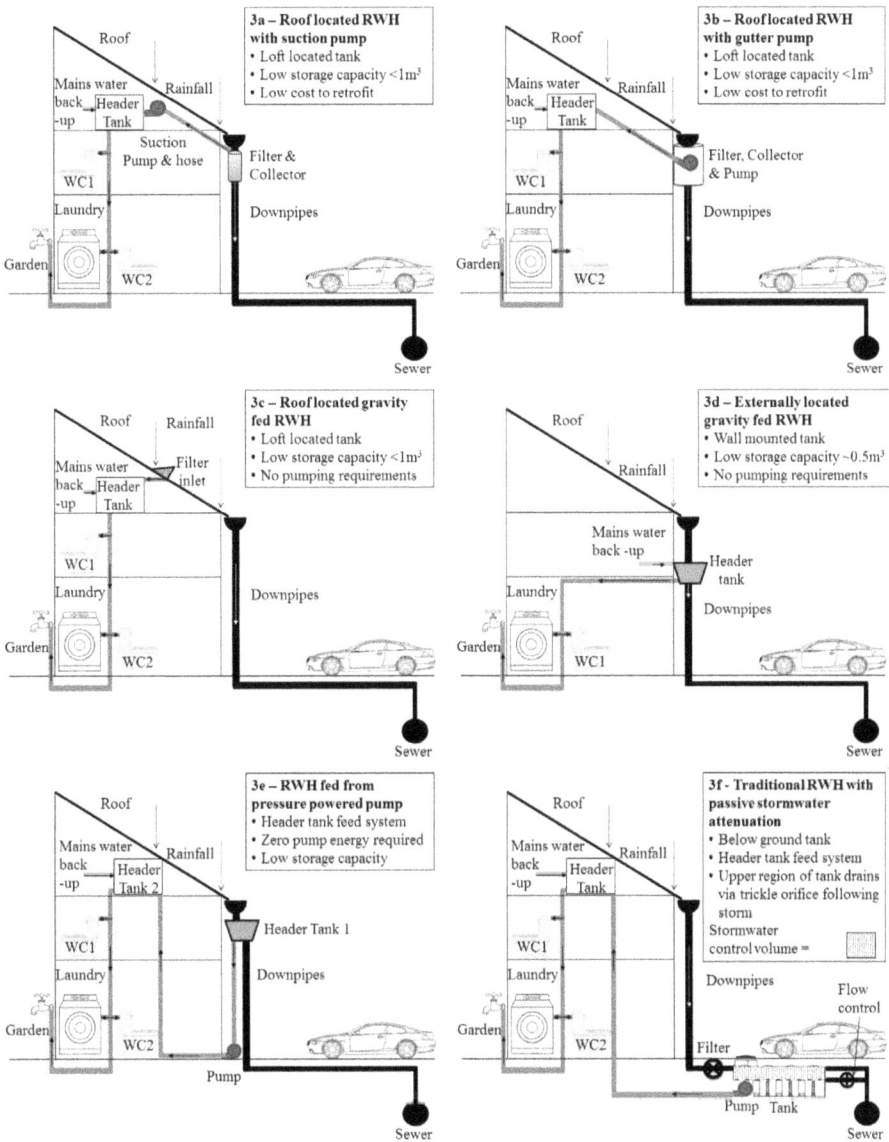

Figure 3. *Cont.*

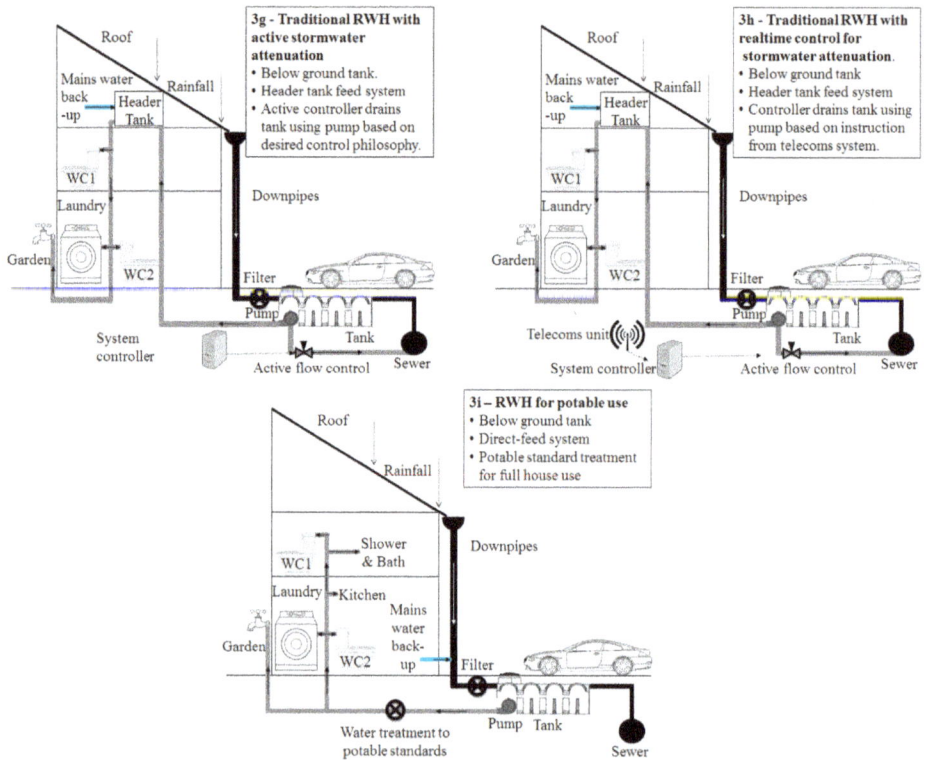

Figure 3. Innovative RWH system configurations emerging in the UK market. (a) Flushrain ; (b)Aqua Harvest and Save; (c) Atlas Water Harvesting / Rooftop Rain; (d) Rainbeetle / Aqualogic Rain Catcher; (e) Hydromentum;(f) RainActiv; (g) KloudKeeper; (h) Real time control RWH; (i) RainSafe.

A common theme with the first five of these innovative RWH system configurations is a high-level roof-runoff inlet, which facilitates the replacement of the large ground-level tank with wall-mounted or internal header tanks. This enables rainwater to be propelled by low energy pumps or flow under gravity into header tanks, which in turn feed appliances by gravity. A second common theme with the next three innovative RWH system configurations (Figure 3f-h) is the inclusion of a "sacrificial" amount of storage that is utilised for stormwater control. These dual-purpose RWH systems enable flow to be released from storage either passively, using an orifice at a specifically designed height in the tank, or actively through a release valve. Figure 3h describes a system that includes functionality to enable a central authority (for example the water service provider (WSP)) to control tank levels based on predictive rainfall, to enable real time control of rainwater discharges to a sewer network. The final innovative RWH system is a treatment train consisting of filtration, UV and ozonation, which is designed to enable harvested rainwater to be treated to potable standards.

3.3. Step 3—Define Criteria and Associated Objectives

Previous RWH evaluation studies have focused on analyses using a traditional set of criteria (capital costs, water saving and energy consumption). In addition to these traditional criteria, this work seeks to investigate emerging criteria associated with stormwater management.

3.3.1. Traditional Criteria: Capital Costs, Water Savings and Energy Consumption

RWH systems are currently installed in the UK to provide alternative water supplies to displace reliance on potable water. Cash savings are generated for homeowners as metered water charges are reduced accordingly [1]. Minimising the capital cost of a system represents the first driver for consideration when appraising RWH configurations. Re-configuring RWH systems to minimise the capital cost could enable the market to further develop by increasing affordability to a larger number of consumer segments. A RWH system's ability to reduce water demand (*i.e.*, contribute to water efficiency) represents the second key criteria when assessing configurations. Reducing the cost of the configuration (perhaps by reducing the tank size and thus storage volume) may reduce the water savings associated with the installation, so assessment of this criterion enables the traditional assessment of RWH benefits to be evaluated. Detailed price comparison information was made available by a UK provider [50]. Where cost data have not been available (for example where prototype systems were identified that have yet to reach the marketplace), estimates were compiled based on component costs and anticipated labour requirements.

Energy consumption associated with the operation of RWH systems has been comprehensively investigated [51]. Roebuck *et al.* [14,17], illustrates the need to monitor and plan for operational energy consumption associated with pumping rainwater in RWH systems. Vieira *et al.* [51], published an extended review of power consumption for RWH systems and drew comparisons against a range of alternative water resources. This research confirmed that theoretical data for median energy consumption (0.20 kWh/m^3) typically underestimate the data from empirical studies (1.40 kWh/m^3). Viera *et al.* [51] also exemplified the challenges associated with generalisations in energy consumption. Factors such as pump efficiency, pipe friction, fittings (e.g., controls such as ball valves), usage rates, pump start-up factors and control systems all have a role to play. In the UK, Ward *et al.* [52], showed that electricity use for a traditional RWH system installed at an office building was 0.54 kWh/m^3. Raw data were also provided by a supplier who monitored their traditional household-scale RWH system (RainDirector) with a header tank feed. This illustrated that it achieved an energy consumption of 0.68 kWh/m^3 in a laboratory setting and that fewer pump starts were needed than for equivalent direct feed systems [53], which would therefore be expected to have a higher consumption.

The mean energy consumption for UK municipal water supply has been reported as 0.60 kWh/m^3 [54]. European average municipal water supplies are also quoted at a similar level of 0.46 kWh/m^3 [54]. RWH configurations that are able to provide water at a lower energy consumption than the municipal supply could therefore be supported on energy/carbon emission reduction grounds. Consequently, the energy consumption of each RWH configuration represents a suitable criterion to review in terms of kWh/m^3 of water delivered [51]. Where possible, energy costs allocated to each RWH system were taken from literature, although some first principles assumptions were necessary (for example RWH systems with lower total head are likely to have a lower energy consumption than those which pump against a higher head).

3.3.2. Emerging Criteria: Stormwater Management and Reducing Combined Sewer Overflows

Through intercepting and using rainwater where it falls (source control), stormwater discharges to sewer systems are reduced as less rainwater enters the sewer network during a storm. Gerolin *et al.* [30], illustrated that RWH can reduce stormwater discharges successfully when the non-potable demand of a property exceeds the rainwater yield. Supporting this, a number of modelling studies on RWH systems have demonstrated their ability to reduce stormwater runoff volumes and rates [55–58]. However, none of the traditional systems outlined in Figure 2 is designed to focus on this functionality.

Variability in extreme rainfall events has been evaluated by Lash *et al.* [59]. This study incorporated modelling approaches (via a probabilistic tank-sizing tool) applied to case study locations in the UK using UK Climate Projections 2009 data. Analysis revealed tank sizes would need to be larger in order to accommodate the increased likelihood of periods with no rainfall. This approach would add support to historic stormwater control approaches set out in Gerolin *et al.* [30], which calls for intentionally oversized RWH tanks to minimise stormwater discharges. The original British Standard for RWH, BS8515:2009 [60], included an approach that encouraged users to size storage tanks to supply a household's non-potable water demand for 18 days. The Standard's "design methods" did not include parameters relating to stormwater control. This single objective approach potentially discouraged technological innovation from RWH system suppliers. Despite this, some technological innovation has been achieved as systems have become increasingly easy to install due to the "plug and play" nature of the components provided [9]. The original Standard also suggests that designers can implement systems that achieve stormwater control by including: *"green roofs; a tank which attenuates flows with an outlet throttle to discharge excess flows; a large tank which is sized for stormwater storage and automatically pumped out or otherwise drained; a tank which is connected to an infiltration system for excess flows."* [60] (p. 32). A recent update to the British Standard [2] now includes an additional technical annex that encourages the design of source control benefits when sizing RWH. However, the stormwater control objective remains outside the scope of the Standard's core tank-sizing calculations. The UK's incumbent RWH system providers do not currently produce systems that provide source control in line with UK guidance. However, it is anticipated that novel configurations that achieve this will be available in the near future as development is underway and products are beginning to be launched [44].

Controlling stormwater discharges to combined sewer networks can mitigate the risks of pollution events from sewage spills during intense rainfall. Reducing combined sewer overflow discharges (in terms of frequency and volume) represents a key area of Asset Management Plan (AMP) investment for a number of UK WSPs [61]. WSP projects such as RainScape, WaterShed and the Urban Demonstrator are underway, which seek to deliver and monitor pilot studies where retrofit stormwater management solutions are being trialled to reduce sewer flooding and associated pollution of watercourses from spills at combined sewer overflows [61–63]. RWH systems that are configured to satisfy the stormwater reductions targeted by WSPs could potentially see them become a viable option alongside other retrofit SuDS approaches over the next decade.

RWH systems evaluated in this study were appraised against two stormwater-related criteria: (1) The reduction in peak daily stormwater discharge volumes; and (2) The reduction in annual average stormwater discharge volumes.

3.3.3. Summarising Criteria for Evaluating RWH Design Configurations

The discussion presented in the previous sections enabled a range of criteria (and associated objectives) to be defined. These are summarised in Table 3 and can be used to evaluate the RWH configurations outlined in Figures 2 and 3. Other criteria, such as the ease of retrofitability, end-user acceptability and lifetime maintenance requirements, have not been considered in the present analysis, but are the subject of on-going research.

Table 3. Criteria for evaluation of RWH system configurations.

Criteria	Objective
C_1 Capital cost of RWH system (£/installation)	O_1 Minimise capital cost of system
C_2 Water Efficiency (m^3/annum potable saved)	O_2 Maximise water saving of system
C_3 Change in operational energy consumption for water supply (kWh/annum)	O_3 Minimise energy required to supply household water
C_4 Reduction in stormwater flow during extreme events (m^3/day)	O_4 Minimise discharge volume of rainwater during largest 24 h storm in 20 year time-series
C_5 Reduction in annual stormwater discharge to sewer (m^3/annum)	O_5 Minimise annual average discharge volume to sewer network

3.4. Step 4—Populating Assessment Matrix: Details of Quantitative Assessment Methods and Criteria for Calculation

In order to populate the assessment matrix, an input/output flow balance model was developed as a VBA spreadsheet tool, based on earlier RWH studies [2,34,64], but here incorporating additional stormwater related outputs. The model uses the "Yield After Spillage" algorithm whereby rainwater is added to the storage volume recorded for the previous time step. Next excess flows are overflowed prior to extracting demand at that time step [64]. Where intentional stormwater discharges are released from either passive or active controls, these also occur prior to demand being extracted [34]. A runoff factor of 0.9 is assumed. A daily time step was used, which matched the 20-year input rainfall time series for Exeter, UK. The model parameters used to define the property and system simulated are given in Table 1. Criteria C_2 (water saving), C_4 (reduction in maximum daily stormwater discharged) and C_5 (reduction in average annual stormwater discharged) were calculated from the flow balance for each RWH system. C_1 (capital cost) and C_3 (change in operational energy consumption) were calculated as explained in the next section. The model enables outputs to be derived from an annual simulation with rainfall, demand and stormwater spill volumes calculated at the daily time step. Outputs were generated for each day of the year, and the simulation was repeated using 20 annual rainfall files.

C_1 **Capital cost of RWH system:** Values for this criterion were derived outside the flow balance model. They were based on best available information on material costs and labour costs required to install each RWH system evaluated. Costs are defined in terms of £/installation and as a percentage of the highest cost option.

C_2 **Water efficiency:** Water efficiency was taken as the average non-potable water demand satisfied by the RWH system over the 20-year simulation period and is given in m^3/annum. The house's remaining potable water demand and the reduction in potable water usage were calculated.

C_3 **Change in operational energy consumption for water supply:** Operational power consumption for each configuration was taken from the literature with first principles used to differentiate between novel systems where empirical data were not available. Power consumption increases and decreases for annual water supply were recorded as a percentage of the baseline scenario (*i.e.*, a house without RWH receiving only municipal water).

C_4 **Reduction in stormwater flow during extreme events:** The largest 24 h storm event recorded over the 20-year period was used to evaluate each RWH system's ability to reduce the discharge volume. The difference between the volume controlled when each RWH system was modelled compared with the volume spilled without RWH was used to define these values. The percentage change between the scenarios was also calculated (*i.e.*, the percentage of stormwater successfully controlled by each RWH system during the largest storm event).

C_5 **Reduction in annual stormwater discharge to sewer:** The annual average reduction in stormwater discharges to the sewer was derived as the difference between uncontrolled overflows to the sewer when RWH is included *vs.* the annual volume spilled without RWH installed. The percentage change between the scenarios was also calculated (*i.e.*, the percentage of annual stormwater successfully controlled by each RWH system).

With simulations completed and outputs derived for the range of options tested, the performance matrix was populated and an analysis conducted as described in Section 4.1.

4. Results and Discussion

4.1. Step 5—Evaluate Performance of RWH Configurations against Criteria

Using the quantitative assessment methods described, an evaluation matrix was defined (Table 4) to summarise simulated performance of the RWH configurations under each criteria.

Table 4. Populated Evaluation Matrix for Criteria Associated With 13 RWH Configurations.

Criteria	No RWH	RWH System (Reference Figure)												
		2a	2b	2c	2d	3a	3b	3c	3d	3e	3f	3g	3h	3i
C_1 Capital cost of RWH system (£)	0	4300	3800	4800	4300	1050	1030	1000	1000	2000	5300	5550	6300	6100
C_1 Capex relative to highest cost option (%)	0	68	60	76	68	17	16	16	16	32	84	88	100	97
C_2 Water Saved (m³/annum)	0	34	34	34	34	23	23	16	23	7	32	34	34	40
C_2 Mains water consumed (m³/annum)	219	185	185	185	185	196	196	202	196	212	187	185	185	179
C_2 Water Efficiency: Change in mains water use/annum (%)	100	84	84	84	84	89	89	92	89	97	85	84	84	82
C_3 Energy cost for RWH system (kWh/m³)	0.0	1.1	1.0	0.8	0.7	0.1	0.1	0.0	0.0	0.0	0.7	0.8	0.9	2.0
C_3 Energy cost of rainwater delivered per annum (kWh)	0	38	35	28	24	3	3	0	0	0	22	28	31	80
C_3 Energy cost of mains water used per annum (kWh)	131	111	111	111	111	117	117	121	117	127	112	111	111	107
C_3 Total energy cost per annum (kWh)	131	148	145	138	134	120	120	121	117	127	135	138	142	187
C_3 Change in operational energy (kWh/annum)	0	17	14	7	3	-11	-11	-9	-13	-4	3	7	10	56
C_3 Change in operational energy (%)	100	113	110	105	103	92	92	92	89	97	102	105	108	143
C_4 Reduction in stormwater flow during extreme event (m³)	0	3.3	3.3	3.3	3.3	0.5	0.5	0.5	0.5	0.5	2.5	3.3	4.9	4.5
C_4 Change in stormwater flow during extreme event (%)	0	67	67	67	67	10	10	10	10	10	51	67	100	92
C_5 Reduction in annual stormwater discharge to sewer (m³/annum)	0	37	37	37	37	22	22	17	22	8	33	37	40	40
C_5 Change in annual stormwater discharge to sewer (%)	0	91	91	93	93	56	56	43	56	20	82	93	100	100

Figure 4 illustrates the ability of each system to perform when plotted against the five criteria. This figure describes the data in a normalised format. For C_1–C_5, each value is divided by the maximum score to give output values between 0 and 1. For C_4 and C_5, this value is subtracted from 1. Hence, the system with a value closest to 0 is the best performing under that criterion. When a single criterion is selected, the results show that there is always at least one novel configuration available that outperforms the four traditional systems (Systems 2a-d). This illustrates that the current RWH configurations deployed in the UK do not necessarily represent the optimal design when broader criteria are included in their evaluation. The traditional RWH systems are outscored by a number of novel RWH system configurations in relation to a number of criteria (for example System 3b and System 3d have lowest cost (C_1) and lowest energy (C_3) ranks). The real time control strategy associated with System 3h was able to fully prevent uncontrolled stormwater spills for every storm in the 20-year study period. In addition, the high demand (600 L/day) associated with System 3.i's potable use ensured this system was also able to reduce the largest storm event in 20 years by 92%. The passive stormwater controls associated with System 3g reduced the extreme storm event by 67% and limited total stormwater discharge volumes to just 7% of the "No RWH" scenario. Evidence from this analysis suggests that the current RWH systems being implemented in the UK can be improved to better satisfy the criteria highlighted in this paper.

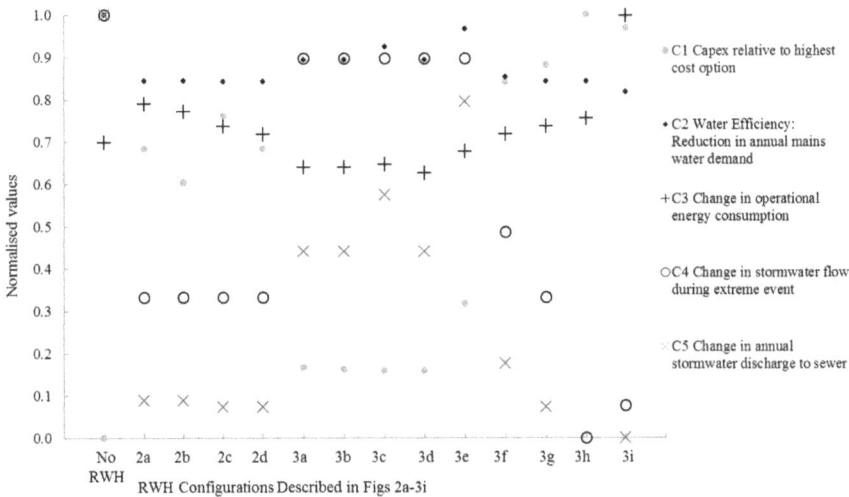

Figure 4. Normalised performance of RWH configurations against a range of design criteria (best scoring systems have lowest values for each criteria).

4.2. Step 6—Scenario Testing

The MCA developed in this paper can be deployed as a method for selecting a configuration for a specific site by adding user-based weightings to define the relative importance of each criterion. Such an approach could be deployed by decision makers to test how different weightings affect the selection of a RWH system. To enable this, the decision maker allocates weightings across each criterion that total 1 unit. Three scenarios are defined below to illustrate in outline how the preferred configuration might be defined by differing decision makers. The assumptions made in the scenarios are based on the authors' knowledge.

Scenario A—A RWH designer concludes that all criteria have equal weight and 0.2 units are applied to each. The MCA suggests that system 3h (traditional RWH with RTC) is the preferred option as it has the lowest total score across all criteria.

Scenario B—A householder wishes to retrofit a RWH system at the lowest possible capital cost. No other criteria hold importance in the system selection process. A weighting of 1 unit is applied to the C_1 *Capex relative to highest cost option* and all other weights set to zero. The MCA selects system 3c3 (roof located, gravity fed gravity RWH) or 3d4 (externally located, gravity fed gravity RWH).

Scenario C—A WSP plans to retrofit houses with RWH as a water demand reduction measure. They also have a secondary objective to reduce peak stormwater flows at a local pumping station. Costs are not an important factor as no alternative solutions have been identified by the WSP. A weighting of 0.75 units is applied to C_2 *Water Efficiency* and 0.25 units allocated to C_4 *Change in stormwater flow during extreme event* to ensure these dominate the remaining criteria. System 3i (RWH for potable use) is identified as dominating other options as this configuration scores best in terms of water demand reduction and is also able to mitigate 92% of the peak storm discharge during the largest storm event tested.

The three scenarios considered above each illustrate the ability of the MCA method to readily offer a high level focus for a designer to identify suitable RWH options under a range of settings/ site objectives.

5. Conclusions

This paper presented the identification, description and multi-criteria analysis of existing and novel RWH configurations that could be adopted at UK households to satisfy a broad range of property and regime level drivers. The evaluation criteria were defined as follows: reduce capital costs, maximise water saving efficiency, minimise operational energy consumption associated with water supply, minimise peak stormwater discharges, and minimise annual stormwater discharges. A broad range of RWH configurations are emerging in the UK marketplace. Through benchmarking each configuration using the MCA, it was possible to score each system's ability to satisfy a number of key RWH criteria. Evidence from the MCA illustrates that the traditional RWH configurations are not necessarily the optimum solutions when broader criteria are considered. However, the specific technology selected will depend on the preferences of the decision maker or user, as illustrated by the three scenarios. Based on these results, it is suggested that minor alterations to existing RWH technologies, such as integration with real time stormwater control devices, could see demand for RWH systems grow in the years ahead. This may be the case where stormwater control is desirable to meet drainage design criteria at new developments, or to reduce sewer flooding and spills in existing combined sewer catchments. The identification of RWH systems as a multi-functional technology is exemplified in this paper. Further empirical studies are now underway to enable novel benefits of emerging RWH system configurations to be further quantified, understood and exploited by a range of decision makers.

Acknowledgments: The study was funded by the UK Engineering & Physical Sciences Research Council supported by Severn Trent Water through delivery of a STREAM Engineering Doctorate. Grant reference: EP/G037094/1. The authors also wish to thank the reviewers for their support in developing this paper.

Author Contributions: All authors conceived the methods set out in this paper and pooled grey and academic literature to enable the review of potential RWH configurations and criteria to be developed. The lead author was responsible for synthesising the configurations and applying the methodology. The co-authors provided significant technical input to ensure the results and discussion were adequately exploited and described herein.

Conflicts of Interest: The work reported here builds upon a series of research and development projects that have been undertaken in collaboration with Severn Trent Water and a number of the RWH system providers referenced in this paper. The authors have sought to balance input from a range of these sources and to minimise the reliance on grey literature when opportunities have permitted.

References

1. Environment Agency. Harvesting Rainwater for Domestic Uses: An Information Guide, 2008. Available online: http://www.highland.gov.uk/NR/rdonlyres/F512E7E7-6C8D-4036-ACFF-1DA761006DF8/0/ RainwaterHarvestingforDomesticUse.pdf (accessed on 10 July 2015).

2. British Standards Institution. *BS 8515:2009+A1:2013—Rainwater Harvesting Systems—Code of Practice*; BSI: London, UK, 2013.

3. Ahmed, W.; Vieritz, A.; Goonetilleke, A.; Gardner, T. Health Risk from the Use of Roof-Harvested Rainwater in Southeast Queensland, Australia, as Potable or Nonpotable Water, Determined Using Quantitative Microbial Risk Assessment. *Appl. Environ. Microbiol.* **2010**, *76*, 7382–7391. [CrossRef] [PubMed]

4. Texas Water Development Board. The Texas Manual on Rainwater Harvesting, 2005. Available online: http://www.twdb.state.tx.us/innovativewater/rainwater/doc/RainwaterCommitteeFinalReport.pdf (accessed on 15 October 2014).

5. Partzsch, L. Smart regulation for water innovation—The case of decentralized rainwater technology. *J. Clean. Prod.* **2009**, *17*, 985–991. [CrossRef]

6. Herrmann, T.; Schmida, U. Rainwater utilisation in Germany: Efficiency, dimensioning, hydraulic and environmental aspects. *Urban Water* **1999**, *1*, 307–316. [CrossRef]

7. Burns, M.J.; Fletcher, T.D.; Duncan, H.P.; Hatt, B.E.; Ladson, A.R.; Walsh, C.J. The performance of rainwater tanks for stormwater retention and water supply at the household scale: An empirical study. *Hydrol. Process.* **2015**, *29*, 152–160. [CrossRef]

8. Stormsaver. Rainwater Harvesting Systems. Available online: http://www.stormsaver.co.uk (accessed on 22 October 2014).

9. Rainwater Harvesting Ltd. Rainwater Harvesting Systems. Available online: http://www.rainwaterharvesting.co.uk (accessed on 12 August 2015).

10. GRAF. Rainwater Harvesting Systems. Available online: http://www.graf-water.com/rainwater-harvesting.html (accessed on 22 May 2015).

11. Ward, S.; Barr, S.; Butler, D.; Memon, F.A. Rainwater harvesting in the UK—Socio-technical theory and practice. *Technol. Forecast. Soc. Chang.* **2012**, *79*, 1354–1361. [CrossRef]

12. Pullinger, M.; Browne, A.L.; Medd, W.; Anderson, B. *Patterns of Practice: Laundry, Bathroom and Gardening Practices of Households in England Influencing Water Consumption and Demand Management*; Lancaster Environment Centre, Lancaster University: Lancaster, UK, 2013; Available online: http://www.sprg.ac.uk/uploads/patterns-of-water-final-report.pdf (accessed on 4 July 2013).

13. MTW Research. *Rainwater Harvesting Market Research & Analysis UK 2010*; MTW Research: Cheltenham, UK, 2010.

14. Roebuck, R.M.; Oltean-Dumbrava, C.; Tait, S. Whole life cost performance of domestic rainwater harvesting systems in the United Kingdom. *Water Environ. J.* **2011**, *25*, 355–365. [CrossRef]

15. Neto, R.F.M.; Carvalho, I.D.; Calijuri, M.L.; Santiago, A.D. Rainwater use in airports: A case study in Brazil. *Resour. Conserv. Recycl.* **2012**, *68*, 36–43. [CrossRef]

16. Ward, S.L.; Memon, F.A.; Butler, D. Rainwater harvesting: Model-based design evaluation. *Water Sci. Technol.* **2010**, *61*, 85–96. [CrossRef] [PubMed]

17. Roebuck, R.M.; Oltean-Dumbrava, C.; Tait, S. Can simplified design methods for domestic rainwater harvesting systems produce realistic water-saving and financial predictions? *Water Environ. J.* **2012**, *26*, 352–360. [CrossRef]

18. Steffen, J.; Jenson, M.; Pomeroy, C.A.; Burian, S.J. Water Supply and Stormwater Management Benefits of Residential Rainwater Harvesting in U.S. Cities. *J. Am. Water Resour. Assoc.* **2013**, *49*, 810–824. [CrossRef]

19. Jones, M.P.; Hunt, W.F. Performance of rainwater harvesting systems in the southeastern United States. *Resour. Conserv. Recycl.* **2010**, *54*, 623–629. [CrossRef]

20. Friedrich, E.; Pillay, S.; Buckley, C.A. The use of LCA in the water industry and the case for an environmental performance indicator. *Water SA* **2007**, *33*, 443–452. [CrossRef]

21. Coombes, P.J. Rainwater Tanks Revisited: New Opportunities for Urban Water Cycle Management. Ph.D. Thesis, University of Newcastle, N.S.W., Australia, 2002. Available online: http://urbanwatercyclesolutions.com/rainwater-tanks-revisited-new-opportunities-for-integrated-water-cycle-management/ (accessed on 1 August 2015).

22. British Standards Institute. *BS8595:2013 Code of Practice for the Selection of Water Reuse Systems*; BSI: London, UK, 2013.

23. Cinelli, M.; Coles, S.R.; Kirwan, K. Analysis of the potentials of multi criteria decision analysis methods to conduct sustainability assessment. *Ecol. Indic.* **2014**, *46*, 138–148. [CrossRef]

24. Urrutiaguer, M.; Lloyd, S.; Lamshed, S. Determining water sensitive urban design project benefits using a multi-criteria assessment tool. *Water Sci. Technol.* **2010**, *61*, 2333–2341. [CrossRef] [PubMed]
25. Hajkowicz, S.; Higgins, A. A comparison of multiple criteria analysis techniques for water resource management. *Eur. J. Oper. Res.* **2008**, *184*, 255–265. [CrossRef]
26. Janssen, R. On the use of multi-criteria analysis in environmental impact assessment in The Netherlands. *J. Multi-Criteria Decis. Anal.* **2001**, *10*, 101–109. [CrossRef]
27. Department for Communities and Central Government (DCLG). *Multi-Criteria Analysis: A Manual*; Communities and Local Government Publications: Wetherby, UK, 2009.
28. Diaper, C.; Jefferson, B.; Parsons, S.A.; Judd, S.J. Water-recycling technologies in the UK. *J. Charter. Inst. Water Environ. Manag.* **2001**, *15*, 282–286. [CrossRef]
29. Lazarova, V.; Hills, S.; Birks, R. Using recycled water for non-potable, urban uses: A review with particular reference to toilet flushing. *Water Recycl. Mediterr. Reg.* **2003**, *3*, 69–77.
30. Gerolin, A.; Kellagher, R.B.; Faram, M.G. Rainwater harvesting systems for stormwater management: Feasibility and sizing considerations for the UK. In Proceedings of the Novatech 2010, Lyon, France, 27 June–1 July 2010.
31. Parsons, D.; Goodhew, S.; Fewkes, A.; De Wilde, P. The perceived barriers to the inclusion of rainwater harvesting systems by UK house building companies. *Urban Water J.* **2010**, *7*, 257–265. [CrossRef]
32. UK Rainwater Harvesting Association (UKRHA). UKRHA-Installers, 2014. Available online: http://www. ukrha.org/tag/system-installers-servicing/ (accessed on 10 October 2014).
33. Butler, D.; Memon, F.A. Water consumption trends and demand forecasting techniques. In *Water Demand Management*; IWA Publishing: London, UK, 2006; pp. 1–26.
34. Roebuck, R.M. A Whole Life Costing Approach for Rainwater Harvesting Systems. An Investigation into the Whole Life Cost Implications of Using Rainwater Harvesting Systems for Non-Potable Applications in New-Build Developments in the UK. Ph.D. Thesis, School of Engineering, Design and Technology University of Bradford, Bradford, UK, 2007.
35. O'Driscoll, N. Rainwater Harvesting System. Patent GB2449534, 8 July 2009.
36. Melville-Shreeve, P.; Ward, S.; Butler, D. A preliminary sustainability assessment of innovative rainwater harvesting for residential properties in the UK. *J. Southeast Univ. Engl. Ed.* **2014**, *30*, 135–142.
37. Woolass, K. Rainwater Collection and Storage. Patent GB2480834, 25 April 2012.
38. Oakley, C. Rainwater Harvesting System. Patent GB2496729, 22 September 2013.
39. Atlas Water Harvesting. Available online: http://www.atlaswaterharvesting.co.uk (accessed on 1 July 2015).
40. Brittain, G. A Rainwater Harvesting System. Patent GB2475924, 21 September 2011.
41. Mottley, R. Roof Tile Rain Collector. Patent GB2228521, 29 August 1990.
42. Bannocks, S. Water Storage Tank. Patent GB2501313, 23 October 2013.
43. Water Powered Technologies. Hydromentum. Available online: http://www.waterpoweredtechnologies. com (accessed on 1 July 2015).
44. Rainwater Harvesting Ltd. Rainwater Harvesting RainActiv. Available online: http://www. rainwaterharvesting.co.uk/downloads/brochures/rain-activ-brochure.pdf (accessed on 12 August 2015).
45. Melville-Shreeve, P.; Ward, S.; Butler, D. Developing a methodology for appraising rainwater harvesting with integrated source control using a case study from south-west England. In Proceedings of the International Conference on Urban Drainage (ICUD), Kuching, Malaysia, 7–12 September 2014.
46. Debusk, K.M.; Hunt, W.F.; Wright, J.D. Characterizing Rainwater Harvesting Performance and Demonstrating Stormwater Management Benefits in the Humid Southeast USA. *JAWRA J. Am. Water Resour. Assoc.* **2013**, *49*, 1398–1411. [CrossRef]
47. Reidy, P. Real-time Monitoring and Active Control of Green Infrastructure: Growing Green Infrastructure. In Proceedings of the 2011 Central New York Green Infrastructure Symposium, New York, NY, USA, 17 November 2011.
48. Han, M.Y.; Mun, J.S. Operational data of the Star City rainwater harvesting system and its role as a climate change adaptation and a social influence. *Water Sci. Technol.* **2011**, *63*, 2796–2801. [CrossRef] [PubMed]
49. RainSafe. Available online: http://www.rainsafe.co.uk (accessed on 1 July 2015).
50. Rainwater Harvesting Ltd. SCP Environmental Ltd.: Installation Options for Rainwater Harvesting Limited. Available online: http://www.rainwaterharvesting.co.uk/downloads/rainwater-harvesting-installation-single-systems.pdf (accessed on 12 August 2015).

51. Vieira, A.S.; Beal, C.D.; Ghisi, E.; Stewart, R.A. Energy intensity of rainwater harvesting systems. *Renew. Sustain. Energy Rev.* **2014**, *34*, 225–242. [CrossRef]
52. Ward, S.; Butler, D.; Memon, F.A. Benchmarking energy consumption and CO_2 emissions from rainwater-harvesting systems: An improved method by proxy. *Water Environ. J.* **2012**, *26*, 184–190. [CrossRef]
53. Rainwater Harvesting Ltd. Rainwater Harvesting: Reduced Use of Electricity by the HydroForce® Pump and Smart Header Tank. Available online: http://www.rainwater-harvesting.co.uk/downloads/pdf/rainwater-harvesting/rwh-white-paper-raindirector-electricity-use.pdf (accessed on 12 August 2015).
54. European Benchmarking Co-Operation. *Learning from International Best Practices 2013*; Water and Wastewater Benchmark: Hamburg, Germany, 2013.
55. Leggett, D.J.; Brown, R.; Stanfield, G.; Brewer, D.; Holliday, E. *Rainwater and Greywater Use in Buildings: Decision-Making for Water Conservation*; CIRIA PR 80; Construction Industry Research and Information Association Publication: London, UK, 2001.
56. Campisano, A.; Cutore, P.; Modica, C.; Nie, L. Reducing inflow to stormwater sewers by the use of domestic rainwater harvesting tanks. In Proceedings of the Novatech 2013, Lyon, France, 23–27 June 2013.
57. Memon, F.A.; Fidar, A.; Lobban, A.; Djordjevic, S.; Butler, D. Effectiveness of rainwater harvesting as stormwater management option. In Water Engineering for a Sustainable Environment, Proceedings of 33rd IAHR Congress, Vancouver, BC, Canada, 9–14 August 2009.
58. Debusk, K.M.; Hunt, W.F. *Rainwater Harvesting: A Comprehensive Review of Literature*; Report 425; Water Resources Research Institute of the University of North Carolina: Chapel Hill, NC, USA, 2014; Available online: http://www.lib.ncsu.edu/resolver/1840.4/8170 (accessed on 10 March 2015).
59. Lash, D.; Ward, S.; Kershaw, T.; Butler, D.; Eames, M. Robust rainwater harvesting: Probabilistic tank sizing for climate change adaptation. *J. Water Clim. Chang.* **2014**, *5*, 526–539. [CrossRef]
60. British Standards Institution. *BS 8515:2009 Rainwater Harvesting Systems—Code of Practice*; BSI: London, UK, 2009.
61. Welsh Water. RainScape Llanelli, 2015. Available online: http://www.dwrcymru.com/en/My-Wastewater/RainScape/RainScape-Llanelli.aspx (accessed on 15 September 2015).
62. South West Water (SWW). Downstream Thinking, 2014. Available online: http://www.southwestwater.co.uk/index.cfm?articleid=11492 (accessed on 10 September 2015).
63. Brewington, J. UK Urban Demonstrator Birmingham, Severn Trent Water, 25 June 2015. Available online: http://www.sustainabilitywestmidlands.org.uk/wp-content/uploads/Urban-Demonstrator-Tyseley-Birmingham-Science-City.pdf (accessed on 15 September 2015).
64. Fewkes, A.; Butler, D. Simulating the performance of rainwater collection and reuse systems using behavioural models. *Build. Serv. Eng. Res.Technol.* **2000**, *21*, 99–106. [CrossRef]

water

MDPI

Article

Assessing Marginalized Communities in Mexico for Implementation of Rainwater Catchment Systems

Gerardo Sámano-Romero *, Marina Mautner, Alma Chávez-Mejía and Blanca Jiménez-Cisneros

Instituto de Ingeniería, UNAM, Circuito Escolar s/n, Delegación Coyoacán, Ciudad de Mexico, D.F. CP 04510, Mexico; MMautner@iingen.unam.mx (M.M.); AChavezM@iingen.unam.mx (A.C.-M.); BJimenezC@iingen.unam.mx (B.J.-C.)
* Correspondence: GSamanoR@iingen.unam.mx; Tel.: +52-55-5623-3600 (ext. 8682); Fax: +52-55-5623-3600 (ext. 8055)

Academic Editor: Ataur Rahman
Received: 19 September 2015; Accepted: 11 December 2015; Published: 8 April 2016

Abstract: Mexico contains a high percentage of marginalized communities, as well as geographic areas with high annual precipitation (approximately 2000 mm). This study uses regional water access and precipitation data to determine municipalities that would most benefit from the installation of Domestic Rain Water Harvesting Systems (DRWHS). The main objective was to find a relationship between local conditions (marginalization, expected level of service, and precipitation) and the physical components of DRWHS. First, monthly precipitation and the number of inhabitants per household were determined for each municipality. Catchment area and tank size were then calculated for a single dwelling by municipality using water demand, run-off coefficient, monthly precipitation, and number of inhabitants per household. In general, municipalities with very low access to municipal water and very high precipitation were found in the southern area of the country. A curve that estimates catchment area based on annual precipitation was developed using the selected municipalities, which produced an average catchment area of 113.3 m^2 for a water demand of 100 L/capita/day. While any DRWHS must be designed specific to local conditions, this study has determined that a regional approximation can be used to select ideal communities for these systems, which can in turn facilitate national implementation.

Keywords: rainwater; Mexico; catchment; domestic supply; water demand; marginalized communities; design

---

1. Introduction

This research assesses the viability of Domestic Rain Water Harvesting Systems (DRWHS) as a water supply method in marginalized communities in Mexico. By focusing on the most vulnerable populations, this research can be used to benefit communities that have the highest need for an alternative water supply system. Access to and quantity of domestic water, marginalization and precipitation were analyzed in Mexico's 2457 municipalities, as were number of houses and inhabitants.

DRWHS technology, which provides water directly to households and enables a number of small-scale productive activities, has the potential to supply water in rural and peri-urban areas where conventional technologies are not feasible [1]. The DRWHS technique is a local intervention with primarily local benefits on ecosystems and human livelihoods, though catchment scale benefits have been modeled for urban systems [2]. The implementation of rainwater harvesting has been increasing as an alternative to conventional methods to reduce the number of people without access to drinking water, especially in rural areas [3].

The Mexican National Water Plan (PNH) has set up funding mechanisms for achieving stable water supplies across the country, specifically in marginalized communities; however, the means that achieving this goal has not been strictly defined [4]. This technical analysis aims to show that DRWHS is a feasible option for municipal and national decision-makers as a sole water source in marginal communities by providing the design parameters necessary to provide water year round. In Mexico marginalization is measured by the National Population Council (CONAPO) every five years using the Marginalization Index. The data from the 2010 CONAPO report was used for the development of the calculations in this study [5].

The use of rainwater catchment and storage systems has an historical importance all over the world and especially in the Pre-Columbian civilizations of Mexico and Central America [6,7]. Since colonization, this practice has been abandoned and replaced with the techniques used in Europe at that time, although rainwater harvesting has recently been "rediscovered".

Liaw and Chiang [8] propose that rainwater harvesting systems can be used for the following purposes:

1. The main source of potable water;
2. A supplementary source of potable water; or
3. A supplementary source of non-potable water.

This research was conducted with the aim of proposing rainwater catchment systems as the main source of potable water according to the standards established by the World Health Organization [9]. These standards are based on the quantity of water delivered and used for households and the requirements of water for health-related purposes to derive a figure of an acceptable minimum to meet the needs for consumption (hydration and food preparation) and basic hygiene.

In Mexico approximately nine million people lack access to safe water [3], with most of them living in scattered communities with high levels of marginalization and isolation due the mountainous nature of the country. Some of these communities also have the highest amount of rain in the whole country.

In general, DRWHS design cannot be standardized because the amount of rainwater provided depends on the local climatic conditions which means that the most important factor relating to the efficiency of a rainwater system is the correct sizing of the rainwater tank [10], although a comparison between estimated and actual performance in large buildings [11] exemplifies that factors, such as catchment size and actual occupancy level, have great impact on tank size and have to be taken into account in order to build confidence in the performance of these systems. In this study, the relationship was explored between the location of the community, the rain levels reported and the marginalization and lack of water distribution systems.

Water availability in Mexico varies significantly both geographically and temporally. Mexico contains nearly 1500 river basins, with most of the surface water concentrated in the southern half of the country. These basins vary greatly in size: 87% of the country's surface runoff is concentrated in 50 basins, and two thirds is concentrated in just seven basins. Rainfall is also concentrated in the southern half of the country, and 68 percent falls between the months of June and September [12]. In 1997, the country was divided into 13 hydrological-administrative regions (RHA) with the objective of organizing the management and preservation of the nation's waters (these hydrological-administrative regions are composed of river basins, but have been created by taking into account municipal borders so as to facilitate the integration of socioeconomic information).

In Mexico municipalities are responsible for providing water and sanitation services; therefore, due to the uneven distribution of water and economic resources among these entities water and sanitation services are also unevenly distributed. While residents in most municipalities in the northern and driest part of the country have regular access to drinking water, in the southern part of the country, the percentage of the population with access to water drops steeply, despite the fact that the vast majority of water resources are concentrated in this part of the country. Due to the lack of access to drinking water in scattered communities in the mountainous and rainy parts of Mexico, DRWHS has the potential to bring public benefits for these communities.

Experience using DRWHS varies largely from country to country, from local initiatives by users or fiscal incentives for new buildings to large national programs or even programs in small islands lacking other water resources [1,13–20]. In this case, this research aims to establish a methodology that helps to identify municipalities with the greatest potential for using this technology and promote further implementation of national public policies in support of communities lacking water services.

2. Methodology

This study assumes that estimated water demand should be determined according to the Levels of Service established by the World Health Organization [9]. Given that precipitation is variable in time and place, the main objective was to find a relationship between local conditions (marginalization, expected level of service, and precipitation) and the physical components of DRWHS, particularly the catchment surface area and the sizing of the water tank.

Therefore, this study assumes that communities that would receive the most benefits from DRWHS should be determined by their marginalization and local rainfall, considering the unreliability of their water resources. As shown in the Figure 1 marginalization coincides with high precipitation levels.

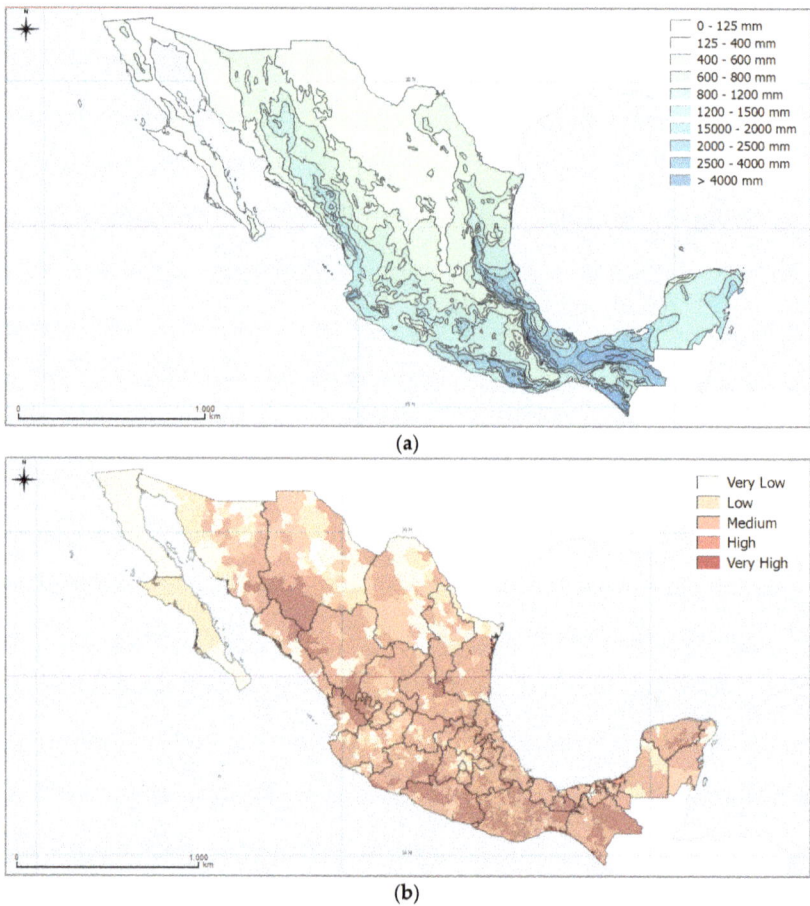

(a)

(b)

Figure 1. (**a**) Average rainfall distribution (source: National Water Commission (CONAGUA)); and (**b**) marginalization in Mexico 2010 index (source: National Population Council (CONAPO)).

The municipality was chosen as the basic unit of study due legal, administrative, and data considerations. Specifically, the supply of water services in Mexico is the responsibility of these entities and most official data, including hydrological data, is indexed by municipality.

2.1. Population and Sample

Starting with the 2457 municipalities that make up the whole of Mexico, various filters were applied to the data to focus in on study areas with the lowest access to drinking water. First, those municipalities with "high" or "very high" degrees of marginalization were chosen as determined by CONAPO. Secondly, of those municipalities, those with 40% or more of the population without access to municipal water were selected [5]. The resulting sample consists of 207 municipalities in 16 states. Annual average rainfall in millimeters was then determined and recorded for each municipality using data from the Institute of Statistic and Geography (INEGI) [21].

Once constructed, the sample was stratified based on the quartile values of two variables: "Access", corresponding to population without access to drinking water, and "Precipitation" (Table 1). Each variable was divided into levels based on these quartiles: Low Access and Very Low Access for Access Level and Very Low, Low, High, and Very High for Precipitation Level. Using these variables, the data set was separated into 8 categories (Figure 2).

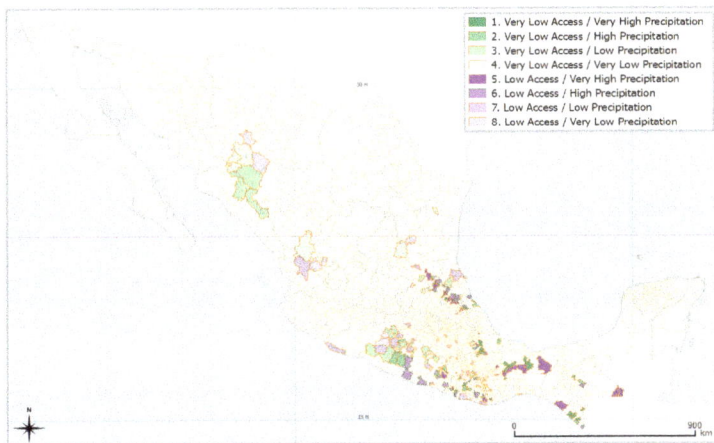

Figure 2. Distribution of the municipalities of the sample.

Table 1. Statistical characteristics of the sample.

Quartile	Population without Municipal Water (%)	Access Level	Precipitation (mm)	Precipitation Level
First	40–45.97	Low	550–900	Very Low
Second	>45.97–53.5		900–1500	Low
Third	>53.5–66.1	Very Low	150–2000	High
Fourth	>66.1–99.7		200–4000	Very High

Figure 2 shows the distribution of the municipalities in the sample. In general, the largest municipalities by area are located in the northern part of the country, while municipalities in the southern part of the country are much more densely situated. The municipalities were color coded from darkest to lightest corresponding to higher to lower level of precipitation. Additionally, the two levels of access represented in the sample were colored green or purple according to Very Low or Low access, respectively. It is apparent that the vast majority of municipalities in the sample are located in the southern portion of the country, which also coincides with the highest amount of rainfall. It was also noted that the high precipitation municipalities in the sample tend to be clustered around the Sierra Madre throughout Mexico, as expected due to its orographic effect which is known to cause precipitation in the surrounding valleys [22].

Of the sample, Category 1 (Very Low Access and Very High Precipitation) was chosen for further analysis. This category represents the greatest potential with the highest impact for capture, purification, and domestic use of rainwater given the shortfall in service and the rainfall regime in those municipalities. The majority of the municipalities in Category 1 are found in the Southern Border and Central Gulf RHAs. All the municipalities are categorized as having more than 50% of the population without access to municipal water and over 2000 mm of rain annually; the range of access to water and precipitation are shown in Figure 3.

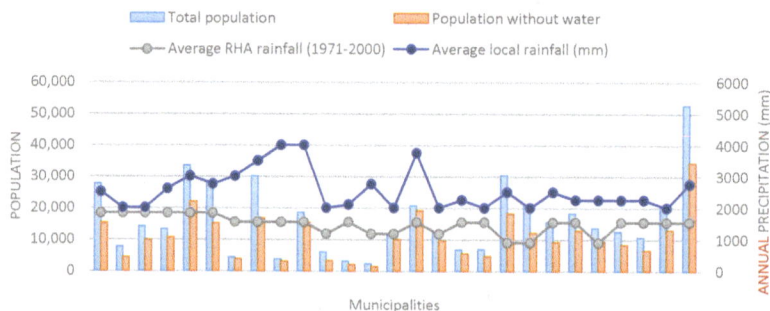

Figure 3. Rainfall distribution and inhabitants without access to drinking water.

This category includes 27 municipalities, most of them in the southern part of the country. The total population in the sample without access to drinking water is 292,752, which calculates to 66% of the population in these municipalities.

2.2. DRWHS Analysis

The physical components of a DRWHS for a single dwelling in each of the municipalities in Category 1 were then determined using the following parameters: water demand, run-off coefficient, monthly precipitation, and number of inhabitants per household. Monthly precipitation and the number of inhabitants per household were determined specific to each municipality, while the water demand and run-off coefficient values were kept constant across all municipalities.

2.2.1. Parameters

Water Demand

Three distinct levels of water demand were chosen to demonstrate how catchment area varies with varying levels of service. Basic Access, at 20 L/capita/day (L/c/d), assures a high health impact by only meeting consumption and basic hygiene needs. Intermediate Access, at 50 L/capita/day, assures a low health impact by meeting consumption, food preparation and hygiene, and personal hygiene needs. Finally, Optimal Access, at 100 L/capita/day, assures no health impact by meeting all hygiene and consumption needs.

Run-off Coefficient

In order to account for variability in catchment surface materials and losses due to absorption and evaporation of rainwater, a run-off coefficient of 80% was considered in all cases. This value was chosen based on a review of literature pertaining to the run-off coefficient [23–25].

Number of Inhabitants per Household

Precise data on bed spaces or average inhabitants per household by municipality was not available on a national level, therefore, the average number of inhabitants per household was determined by dividing the number of residents by the number of households in each municipality. The number of inhabitants, the number of households, and the number of houses without connection to municipal water in each municipality were obtained from the National Municipal Information System (SNIM) [26].

Monthly Precipitation

The monthly rainfall averages were obtained from the official repository of climatological data [27], which is managed by the National Water Commission (CONAGUA) and the National Meteorological Service (SMN). This data is derived from daily precipitation data from individual climatological stations across the country and has been processed and validated by CONAGUA. There has not been a similar processing and validation effort realized by CONAGUA for daily rainfall averages; therefore, the monthly values were considered the most accurate representation of local conditions available. For the purposes of the calculations, the values used correspond to the period 1981–2010, the most recent period of validated averages of 30 years available.

To obtain an average precipitation corresponding to each municipality the Thiessen polygons method [28] was used with the territory of each municipality used instead of the watershed as the geographic limit. With this method a weighted average is calculated from values belonging to polygons intersecting a given sub-region of the municipality. This can be expressed mathematically as:

$$\overline{Ap_l} = \frac{1}{Ms} \sum_{i=1}^{n} CS_a \times Ap_i = \sum_{i=1}^{n} \%Th_i Ppn_i \tag{1}$$

where Ms represents the total area of the municipality, CS_a the area of the municipality sub-region corresponding to a given climatological station, and Ap_i the average precipitation in each month. The concept of $\%Th_i$ is equivalent to the quotient between CS_a and Ms. Each variable was obtained using QGIS software as shown in Figure 4.

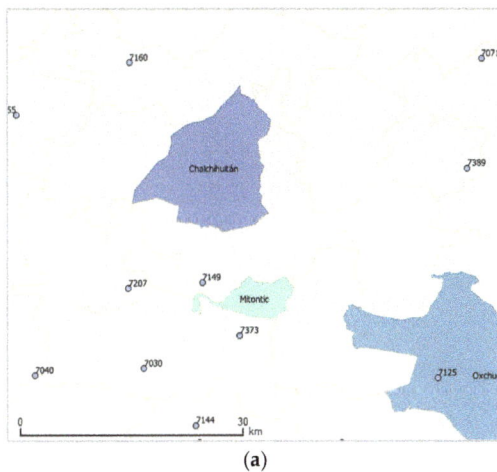

(a)

Figure 4. *Cont.*

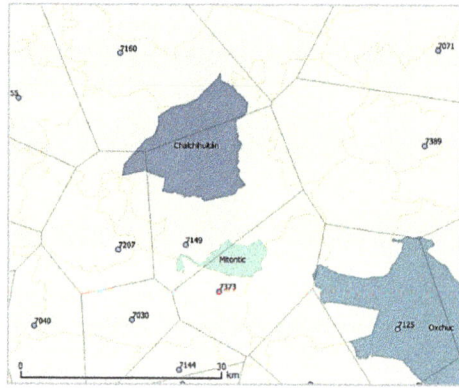

(b)

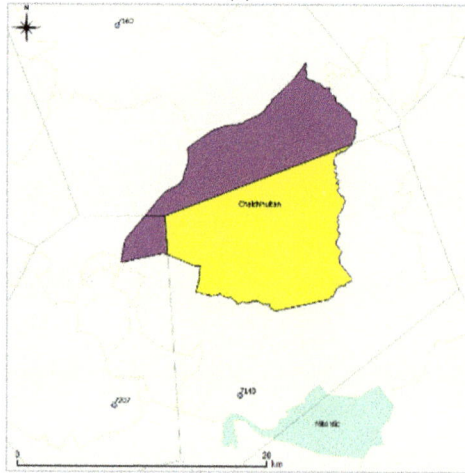

(c)

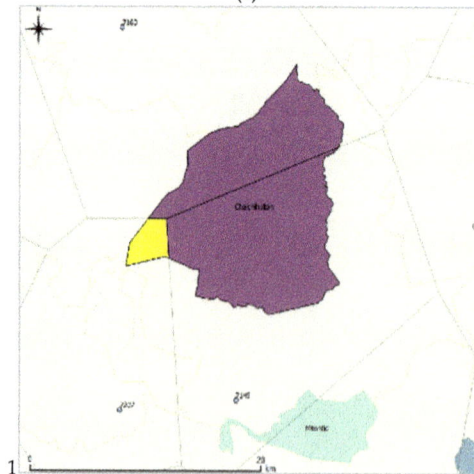

(d)

Figure 4. *Cont.*

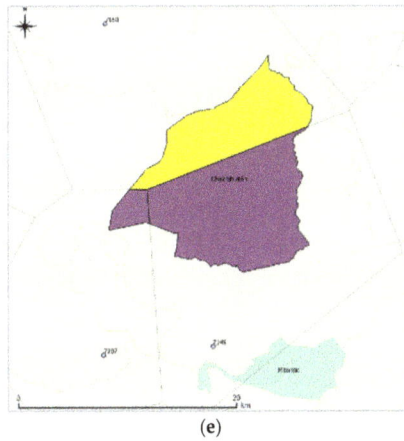

(e)

Figure 4. Example of the Thiessen polygons method applied to the municipal analysis. (**a**) Identification of the municipalities and the nearest climatological stations with normalized data from 1981–2010; (**b**) construction of the Thiessen polygons around each station; (**c**) identification of the influence of the station 7149; (**d**) identification of the influence of the station 7207; and (**e**) identification of the influence of the station 7160.

2.2.2. Catchment Surface Area

Due to heterogeneity in the types and sizes of the houses, catchment area is treated as variable. In each of the municipalities a starting catchment area was estimated based on the approximate annual rainfall. This estimate was then raised or lowered by 1 m² and 5 m² at a time depending on the level of service (Basic and Intermediate/Optimal, respectively) until water demand for the year and average household size was met, resulting in water catchment areas with a precision of ±1 m² and ±5 m² (Basic and Intermediate/Optimal, respectively).

2.2.3. Tank Sizing

To determine the size of the water tank the mass curve method, a critical period method, was used [29]. A critical period method analyzes the absolute difference between demand and supply in sequences of flows to determine the storage capacity, with the mass curve method being the best known and earliest example of this approach to storage sizing [8]. Storage tank modeling has been assessed by various studies using hourly, daily, and monthly data as well as both behavioral and critical period models; a good review of the literature is found in Fewkes and Butler 2000, which compares a variety of these models. In this case monthly data was used, as Fewkes and Butler have shown that monthly data can be used to model the performance of large stores, specifically when demand is equal to 100% or more of annual harvestable rainfall as is the case in this study [30]. The storage change in a tank is calculated using a mass balance equation:

$$I_i = \overline{Ap_l} \times R_q \times C_s \tag{2}$$

where I_i is the inflow of the month analyzed and equal to the product of the average rainfall $\overline{Ap_l}$, the run-off quotient R_q, and the catchment area C_s. As discussed above, C_s was calculated using various levels of water demand.

The outflow in each month corresponds to the product of the number of inhabitants α, the water demand W_d (level of service: 20, 50, or 100 L/capita/day) and the number of days in the month analyzed m_{days}:

$$O_i = \alpha \times W_d \times m_{days} \tag{3}$$

The respective values of inflow and outflow were then summed over all months starting in the month with the highest rainfall. Lastly, the accumulated outflow was subtracted from the accumulated inflow. Therefore, the volume is equal to the sum of the maximum positive difference (the surplus of water that must to be stored in a given month) plus the absolute value of the maximum negative difference (the shortage of water in a given month), which can be expressed mathematically as:

$$Wt_V = max_{(+)}(I_a - O_a) + \left| max_{(-)}(I_a - O_a) \right| \tag{4}$$

where Wt_V is the estimated water tank volume, I_a is the accumulated inflow, and O_a is the accumulated outflow. Figure 5 shows an example of the results of these calculations.

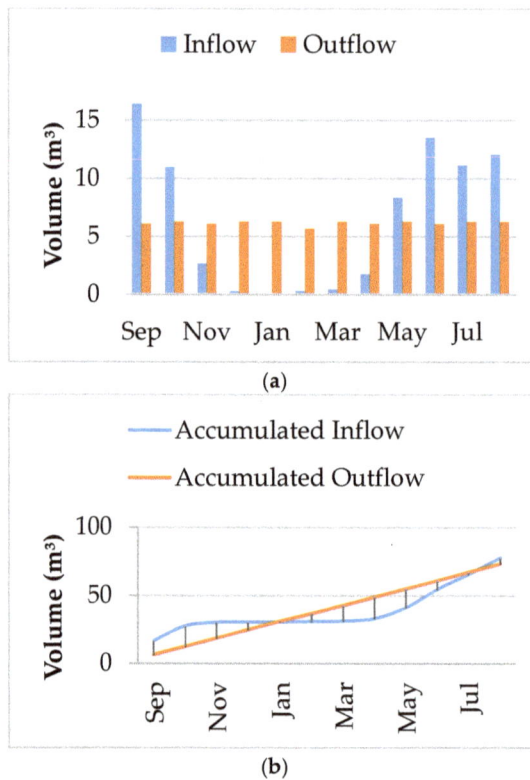

(a)

(b)

Figure 5. Tank sizing analysis in Huehuetan municipality southern Mexico (a) monthly differences between inflow and outflow; and (b) mass curve method applied to the sizing of the water tank.

3. Results and Discussion

Ideal catchment surface area and approximate rainwater tank dimensions were analyzed by comparing the correlation with rainfall annual averages. Both design variables were taken as a function of the rainwater demand established by the level of service.

Table 2 shows the results for area and volume obtained for each of the 27 municipalities analyzed. Slope and coefficient of correlation of the relationship between demand and catchment area are shown in the last columns in Table 2, it is apparent that a linear relationship exists between the required area and the corresponding access level (*i.e.*, from 20 L/capita/day to 50 L/capita/day the catchment area is multiplied by 2.5).

Table 2. Rainfall calculated, Catchment Surface Area and Storage Tank Size required to meet Basic, Intermediate, and Optimal Access.

State	Municipality	Houses		Inhabitants per Household	Calculated Average Annual Rainfall (mm)	Catchment Surface Area and Storage Tank Size						M Area	R^2 Area
		Total	Without Access			Basic Access		Intermediate Access		Optimal Access			
						Area (m^2)	Vol (m^3)	Area (m^2)	Vol (m^3)	Area (m^2)	Vol (m^3)		
Chiapas	Acapetahua	6909	3668	3.9	1837.3	20	12	50	30	100	61	4.00	1.00
	Bejucal de Ocampo	1225	689	5.8	1375.9	40	17	100	42	200	84	2.00	1.00
	Chalchihuitán	8060	5224	4.2	1744.7	22	9	55	23	110	45	1.60	1.00
	El Porvenir	2182	1713	6.1	1788.6	32	16	80	40	160	80	1.60	1.00
	Huehuetán	8060	5224	4.2	2420.5	16	12	40	31	80	62	1.48	1.00
	Villa Comaltitlán	6607	3498	4.2	1574.3	25	12	60	31	120	62	1.40	1.00
Oaxaca	Eloxochitlán de F. M.	1061	902	4.0	3115.0	12	11	30	27	60	53	1.19	1.00
	Huautla de Jiménez	7300	4056	4.1	2300.8	16	11	40	28	80	57	1.20	1.00
	San José Independencia	894	708	4.1	4162.7	10	8	25	20	50	39	1.08	1.00
	San José Tenango	4434	3646	4.2	4162.7	10	8	25	20	50	39	1.20	1.00
	San Lorenzo	1352	756	4.4	1579.6	24	14	60	36	120	72	1.20	1.00
	San Lucas Camotlán	651	415	4.7	598.7	80	23	200	58	400	115	1.10	1.00
	San Mateo Piñas	599	340	3.7	1291.5	32	13	75	32	150	63	0.88	1.00
	San Pedro Jicayán	2344	1987	4.9	1982.2	24	18	60	46	120	92	0.88	1.00
	Santa María Chilchotla	4842	4439	4.3	4239.3	10	9	25	23	50	47	1.00	1.00
	Santiago Amoltepec	2415	1860	5.1	1493.5	32	16	80	41	160	81	1.00	1.00
	Santiago Yaveo	1632	1316	4.1	2160.8	19	10	45	26	90	52	0.89	1.00
Puebla	Tenampulco	1866	1236	3.6	1712.3	24	7	55	17	110	33	0.89	1.00
San Luis Potosí	Matlapa	6638	3847	4.6	1897.8	24	12	60	29	120	58	0.68	1.00
	Tancanhuitz	4678	2601	4.5	1664.9	28	12	70	29	140	59	0.68	1.00
Veracruz	Coxquihui	3488	2081	4.4	2204.7	19	6.4	45	16	90	32	0.80	1.00
	Hidalgotitlán	4466	3109	4.1	2600.0	16	8.2	35	21	70	41	0.80	1.00
	Ilamatlán	3332	2238	4.1	2109.4	20	9.5	45	24	90	48	0.80	1.00
	Santiago Sochiapan	2570	1951	4.2	2381.1	16	9.7	40	24	80	48	0.60	1.00
	Texcatepec	2258	1443	4.5	1961.4	20	9.5	50	24	100	47	0.50	1.00
	Texistepec	5109	3242	4.0	2191.8	20	8.6	45	21.5	90	43	0.50	1.00
	Tzonapa	13,271	8273	4.0	2681.7	16	10	35	25	70	51	0.50	1.00

It is important to note that the calculated annual rainfall in Table 2, is different than the data reported by INEGI. Most of the municipalities have high calculated precipitation, around 2000 mm, with the exception of San Lucas Camotlán, approximately 600 mm. This discrepancy is likely due to inconsistency between the national datasets used by INEGI and the climatological data from CONAGUA used to calculate the average rainfall.

To achieve Optimal Access the required catchment area among the Category 1 municipalities ranges from approximately 60 to 160 m^2 with an average of 113.3 m^2. Using the linear relationship established between water demand and the catchment area, an average of 56.7 and 23.2 m^2 would be required to meet Intermediate and Basic Access respectively. These catchment areas are higher than catchment areas that have been reported [31]. Similarly, for Optimal Access, tank size varied from 32.1 to 115 m^3 with an average of 58.0 m^3 and Intermediate and Basic Access equated to an average size of 29.0–11.6 m^3, respectively. Again, the tank sizes determined in this study are higher than reported in the literature [25].

The difference in the design parameter values reported in other studies and those determined in this study can be attributed to the intentions of this study as compared with others. While other studies focus on the ability of DRWHS to provide a supplement to other water supplies available, this study analyzes the feasibility of using DRWHS as a sole water source throughout the entire year in regions where other water sources are cost prohibitive. National and municipal decision makers can therefore use the tool developed to estimate regional catchment and storage needs to determine relative costs between DRWHS and other water supply infrastructure (e.g., piped water, water trucks) and prioritize areas in which DRWHS is most practicable.

Additionally, Fewkes and Butler reported that monthly data models can sometimes overestimate storage volume; as such, it is suggested that further research be carried out regarding how the performance of this model varies when using differing data time periods (e.g., hourly, daily) [30]. Finally, as noted above, Mexico experiences the majority of its rainfall June through September with relatively dry periods the rest of the year, except in select areas in the southern part of the country. As such, larger tanks sizes are needed than those needed in regions with relatively consistent precipitation throughout the year.

Municipalities with higher annual rainfall can achieve Optimal Access with smaller areas than those with comparatively lower rainfall as is shown in the models developed. However, it is apparent that rainwater storage tank size does not follow a similar behavior and instead depends more on the frequency or regularity of rainfall throughout the year. The municipalities with well-defined periods of drought, even those with high average annual rainfall, ended up requiring larger tanks to store rainwater than those with more even rainfall from month to month. Broadly, it was observed that all parameters (catchment area, inhabitants, water demand, and rainfall amount) influence the sizing of the rainwater tanks to some extent, but the greatest influence by far is the periodicity of the rainfall.

Correlations

Due the relationship that exists between precipitation and the components of DRWHS it is possible to establish a mathematical model that will allow us to estimate the catchment surface area required to meet the water demand. In this case the model was built for an average of four persons per house and precipitations from just under 1300 mm to slightly more than 4200 mm annually. Figure 6 shows the functions obtained from the experimental results.

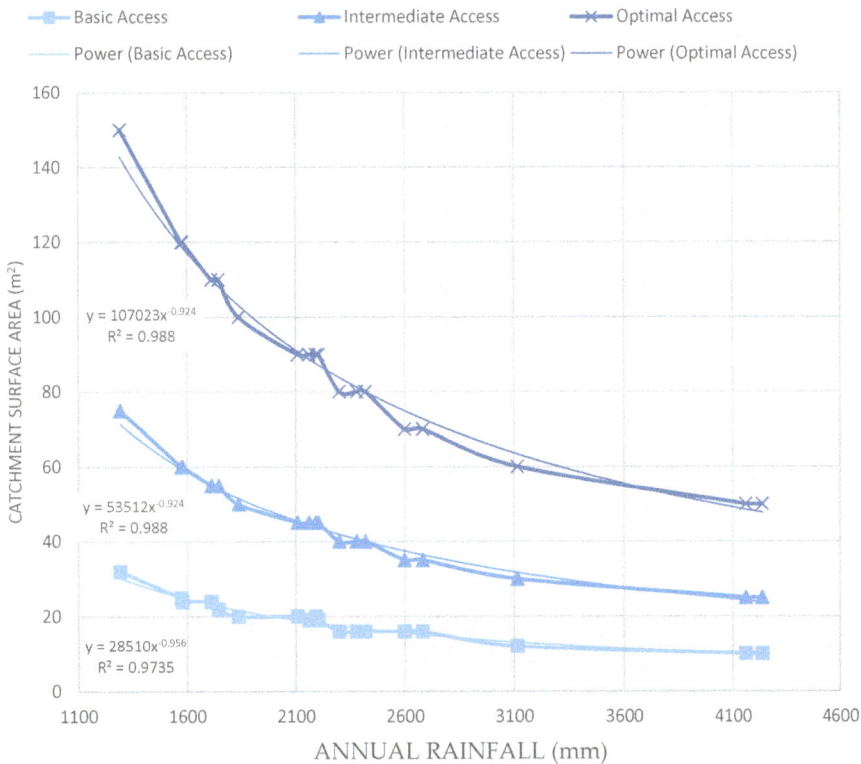

Figure 6. Correlations models for rainfall and catchment surface area.

4. Conclusions

The focus of this study was to evaluate the potential use of rainwater catchment systems in marginalized communities to serve as the sole supply of potable water. The method developed prioritizes areas of high marginalization and high precipitation, while still providing a tool that can be used throughout Mexico for municipal and state level decision-making. High priority communities were selected as those with >40% of the popula tion without access to municipal water sources and 2000 mm annual precipitation due to their potential for high impact from the installation of DRWHS.

Values for catchment area and storage tank size were modeled for the 27 municipalities in the study using average annual rainfall by municipality as determined using the Thiessen polygon method; average number of inhabitants per household as determined by national data; and three different levels of service: Basic (20 L/hab/day), Intermediate (50 L/hab/day), and Optimal (100 L/hab/day), as determined by WHO guidelines.

A linear relationship between rainwater catchment area size and level of service was confirmed with a correlation coefficient of 1.00. A relationship was also found between annual rainfall and catchment area with $R^2 > 0.97$ for all three levels of service, providing a quick method to determine catchment area given level of service and annual rainfall.

Storage tank sizes determined in this study were found to be greater than those reported in the literature which can be attributed to the following influences decreasing in supposed order of importance: the systems were modeled with the intention of serving as the sole supply of potable water as opposed to a supplementary supply; this method employs the use of readily available monthly data as opposed to difficult to obtain daily or hourly data; and much of Mexico experiences the majority of its precipitation during a relatively short four month period as opposed to evenly distributed rainfall throughout the year.

To develop a national rainwater harvesting program in Mexico it is necessary to have a local approach, however, using large scale models as a first approximation will facilitate such a daunting task. No doubt Mexico has the circumstances to exploit the technology of Domestic Rain Water Harvesting Systems given its geographic conditions and its shortcomings in water supply.

Acknowledgments: The authors are thankful for the financial support provided by the Comision Nacional del Agua (formerly CONAGUA) as project sponsor within project 4381, CNA-SGAPDS-CONVENIO-027/2014.

Author Contributions: Gerardo Sámano-Romero and Alma Chávez-Mejía conceived and designed the experiments; Gerardo Sámano-Romero performed the experiments; Gerardo Sámano-Romero and Marina Mautner analyzed the data, Alma Chávez-Mejía and Blanca Jiménez-Cisneros and contributed materials/analysis tools and support; Gerardo Sámano-Romero and Marina Mautner wrote the paper.

Conflicts of Interest: The authors declare no conflict of interest.

References

1. Mwenge-Kahinda, J.-M.; Taigbenu, A.E.; Boroto, J.R. Domestic rainwater harvesting to improve water supply in rural South Africa. *Phys. Chem. Earth* **2007**, *32*, 1050–1057. [CrossRef]
2. Melville-Shreeve, P.; Mugume, S.; Ward, S.; Butler, D. Retrofitting Rainwater Harvesting to Maximise SuDS: A catchment scale assessment of novel RWH configuration. In Proceedings of the Taking SUDS forward, Coventry, UK, 3–4 September 2015.
3. United Nations Environment Programme. *Rainwater Harvesting: A Lifeline for Human Well-being*; A Report Prepared for UNEP by Stockholm Enviroment Institute; United Nations Environment Programme: Nairobi, Kenya, 2009.
4. Comisión Nacional del Agua, National Water Commission. *National Water Program*; National Water Commission, Ed.; Secretaría de Medio Ambiente y Recursos Naturales: Mexico City, Mexico, 2014.
5. Consejo Nacional de Población, National Population Council. *Marginalization Index for Federal Entity and Municipalities*; Consejo Nacional de Población: Mexico City, Mexico, 2011.
6. Comisión Nacional del Agua, National Water Commission. *Historical Sketch of Water in Mexico*; National Water Commission, Ed.; Secretaría de Medio Ambiente y Recursos Naturales: Mexico City, Mexico, 2009.
7. Mays, L.; Antoniou, G.; Angelakis, A. History of water cisterns: Legacies and lessons. *Water* **2013**, *5*, 1916–1940. [CrossRef]
8. Liaw, C.-H.; Chiang, Y.-C. Dimensionless analysis for designing domestic rainwater harvesting systems at the regional level in northern Taiwan. *Water* **2014**, *6*, 3913–3933. [CrossRef]
9. Howard, G.; Bartram, J. *Domestic Water Quantity, Service, Level and Health*; World Health Organization: Geneve, Switzerland, 2003; p. 39.
10. Ghisi, E. Parameters influencing the sizing of rainwater tanks for use in houses. *Water Resour. Manag.* **2010**, *24*, 2381–2403. [CrossRef]
11. Ward, S.; Memon, F.A.; Butler, D. Performance of a large building rainwater harvesting systems. *Water Res.* **2012**, *46*, 5127–5134. [CrossRef] [PubMed]
12. CONAGUA, National Water Commission. *Financing Water Resources Management in Mexico*; Secretaría de Medio Ambiente y Recursos Naturales: Mexico City, Mexico, 2010; p. 32.
13. Hajani, E.; Rahman, A. Reliability and cost analysis of a rainwater harvesting system in peri-urban regions of Greater Sydney, Australia. *Water* **2014**, *6*, 945–960. [CrossRef]
14. Nanninga, T.A.; Bisschops, I.; López, E.; Martínez-Ruiz, J.L.; Murillo, D.; Essl, L.; Starkl, M. Discussion on sustainable water technologies for peri-urban areas of Mexico city: Balancing urbanization and environmental conservation. *Water* **2012**, *4*, 739–758. [CrossRef]
15. Seo, Y.; Park, S.; Kim, Y.-O. Potential benefits from sharing rainwater storages depending on characteristics in demand. *Water* **2015**, *7*, 1013–1029. [CrossRef]
16. Souza, E.L.; Ghisi, E. Potable water savings by using rainwater for non-potable uses in houses. *Water* **2012**, *4*, 607–628. [CrossRef]

17. Sturm, M.; Zimmermann, M.; Schütz, K.; Urban, W.; Hartung, H. Rainwater harvesting as an alternative water resource in rural sites in central northern Namibia. *Phys. Chem. Earth A B C* **2009**, *34*, 776–785. [CrossRef]
18. Gomes, U.F.; Heller, L.; Pena, J. A national program for large scale rainwater harvesting: An individual or public responsibility? *Water Resour. Manag.* **2012**, *26*, 2703–2714. [CrossRef]
19. Dillaha, T.A., III; Zolan, W.J. Rainwater catchment water quality in Micronesia. *Water Res.* **1985**, *19*, 741–746. [CrossRef]
20. Sazakli, E.; Alexopoulos, A.; Leotsinidis, M. Rainwater harvesting, quality assessment and utilization in Kefalonia Island, Greece. *Water Res.* **2007**, *41*, 2039–2047. [CrossRef] [PubMed]
21. Compendio de Información Geográfica Municipal. Available online: http://www.inegi.org.mx/geo/contenidos/topografia/compendio.aspx (accessed on 25 August 2015).
22. Houze, R.A. Orographic effects on precipitating clouds. *Rev. Geophys.* **2012**, *50*. [CrossRef]
23. Farreny, R.; Morales-Pinzón, T.; Guisasola, A.; Taya, C.; Rieradevall, J.; Gabarrell, X. Roof selection for rainwater harvesting: Quantity and quality assessments in Spain. *Water Res.* **2011**, *45*, 3245–3254. [CrossRef] [PubMed]
24. Ghisi, E.; da Fonseca-Tavares, D.; Luis-Rocha, V. Rainwater harvesting in petrol stations in Brasília: Potential for potable water savings and investment feasibility analysis. *Resour. Conserv. Recycl.* **2009**, *54*, 79–85. [CrossRef]
25. Liaw, C.-H.; Tsai, Y.-L. Optimum storage volume of rooftop rain water harvesting systems for domestic use. *JAWRA J. Am. Water Resour. Assoc.* **2004**, *40*, 901–912. [CrossRef]
26. Sistema Nacional de Información Municipal. Available online: http://www.snim.rami.gob.mx/ (accessed on 23 August 2015).
27. CONAGUA, National Water Commission. Climatological Normals. Available online: http://smn.cna.gob.mx/index.php?option=com_content&view=article&id=42:normales-climatologicas-por-estacion&catid=16:general&Itemid=75 (accessed on 17 August 2015).
28. Thiessen, A.H. Precipitation averages for large areas. *Mon. Weather Rev.* **1911**, *39*, 1082–1084. [CrossRef]
29. Unidad de Apoyo Técnico para el Saneamiento Básico del Área Rural. *Design Guidelines for Rainwater Harvesting Systems*; Pan American Center for Sanitary Engineering and Environmental Sciences: Lima, Peru, 2001.
30. Fewkes, A.; Butler, D. Simulating the performance of rainwater collection and reuse systems using behavioural models. *Build. Serv. Eng. Res. Technol.* **2000**, *21*, 99–106. [CrossRef]
31. Abdulla, F.A.; Al-Shareef, A.W. Roof rainwater harvesting systems for household water supply in Jordan. *Desalination* **2009**, *243*, 195–207. [CrossRef]

water

MDPI

Article

A Methodology to Assess and Evaluate Rainwater Harvesting Techniques in (Semi-) Arid Regions

Ammar Adham [1,2,*], Michel Riksen [1], Mohamed Ouessar [3] and Coen J. Ritsema [1]

1   Wageningen University, Soil Physics and Land Management Group, P.O. Box 47, 6700 AA Wageningen,
    The Netherlands; michel.riksen@wur.nl (M.R.); coen.ritsema@wur.nl (C.J.R.)
2   University of Anbar, 31001 Ramadi, Iraq
3   Institut des Régions Arides, Route de Djorf km 22.5, 4119 Medenine, Tunisia; ouessar@yahoo.com
*   Correspondence: ammar.ali@wur.nl or Engammar2000@Yahoo.com; Tel.: +31-659-300-384;
    Fax: +31-317-426-101

Academic Editor: Ataur Rahman
Received: 17 November 2015; Accepted: 4 May 2016; Published: 13 May 2016

Abstract: Arid and semi-arid regions around the world face water scarcity problems due to lack of precipitation and unpredictable rainfall patterns. For thousands of years, rainwater harvesting (RWH) techniques have been applied to cope with water scarcity. Researchers have used many different methodologies for determining suitable sites and techniques for RWH. However, limited attention has been given to the evaluation of RWH structure performance. The aim of this research was to design a scientifically-based, generally applicable methodology to better evaluate the performance of existing RWH techniques in (semi-) arid regions. The methodology integrates engineering, biophysical and socio-economic criteria using the Analytical Hierarchy Process (AHP) supported by the Geographic Information System (GIS). Jessour/Tabias are the most traditional RWH techniques in the Oum Zessar watershed in south-eastern Tunisia, which were used to test this evaluation tool. Fifty-eight RWH locations (14 jessr and 44 tabia) in three main sub-catchments of the watershed were assessed and evaluated. Based on the criteria selected, more than 95% of the assessed sites received low or moderate suitability scores, with only two sites receiving high suitability scores. This integrated methodology, which is highly flexible, saves time and costs, is easy to adapt to different regions and can support designers and decision makers aiming to improve the performance of existing and new RWH sites.

Keywords: RWH suitability; AHP approach; GIS; Tunisia; jessour; tabias

---

1. Introduction

Aridity and climate change are the major challenges faced by farmers who rely on rainfed farming [1]. Especially in arid regions, farmers are faced with low average annual rainfall and variability in temporal and spatial distribution. In order to increase the availability of water for crop production and cattle grazing, inhabitants of dry areas have constructed and developed several types of Rain Water Harvesting techniques (RWH). RWH is a method for inducing, collecting, storing and conserving local surface runoff for agriculture in arid and semi-arid regions [2]. RWH is a likely viable option to increase water productivity at the production system level [3]. RWH and management techniques have a significant potential for improving and sustaining the rainfed agriculture in the region [4]. In fact, a wide variety of micro-catchment, macro-catchment and *in situ* RWH techniques are available in arid and semi-arid regions. The indigenous techniques, or those modified by the indigenous RWH practices, are more common and widely accepted by smallholder farmers than the others [5]. Throughout history, archaeological evidence has revealed RWH sites that were implemented in Jordan, the Al-Negev desert, Syria, Tunisia and Iraq. The earliest signs of RWH are believed to have been constructed over 9000 years ago in the Edom Mountains in southern Jordan [6,7]. The most common RWH techniques in arid and semi-arid regions are dams, terracing, ponds and pans, percolation tanks and Nala bunds. Tunisia is an example of the Mediterranean countries that are facing scarcity of water which will be worsened due to climate change, growing demand for water in agricultural and urban development and an expanding tourism industry [8]. To adapt to this development, Tunisians have developed and implemented several types of water harvesting techniques of which the most common are jessour, tabias, terraces, cisterns, recharge wells, gabion check dams and mescats [9,10].

The success of RWH systems depends mainly on identification of suitable sites and technologies for the particular area. Soil Conservation Service (SCS) with Curve Number (CN), Geographic Information System (GIS) and Remote Sensing (RS) and integrated GIS, RS with Multi-Criteria Analysis (MCA), have all been applied with different biophysical and socio-economics criteria to identify suitable locations for RWH. Several researchers have presented and applied the SCS with the CN method to assess how much runoff can be generated from a runoff area like in South Africa [11], and India [12,13].

Nowadays, the Geographic Information System and Remote Sensing are used to represent the biophysical environment and applied to identify suitable sites for RWH [1,10,14]. Other researchers have integrated GIS, RS and Multi-Criteria Analysis to assess the suitability of sites for RWH [15,16].

Ouessar *et al.* [17] developed and applied a simple tool to evaluate the structural stability of 12 sites (four jessour, four tabias and four gabion check dams) in southern Tunisia. Through physical inspection, the characteristics of the structures were rated and an overall score was given. The characteristics rated include a cross-section for the water and sediment components of the structure, infiltration potential, vegetation quantity, dyke material and dyke erosion. This study also assessed the hydrological impact of the water harvesting systems by adaptation and evaluation of the soil and water assessment model (SWAT).

Jothiprakash and Mandar [18] applied the Analytical Hierarchy Process to evaluate various RWH techniques (aquifer recharge, surface storage structures and concrete storage structures) in order to identify the most appropriate technique and the required number of structures to meet the daily water demand of a large-scale industrial area.

So far, most attention has been given to the selection of suitable sites and techniques for RWH [19] but little attention has been given to the evaluation of the RWH structure after implementation.

To understand the performance of RWH and to ensure successful implementation of new RWH, engineering (technical), biophysical and socio-economic criteria need to be integrated into the evaluation tools [20,21]. In addition, the relation and importance of the various criteria also needs to be taken into consideration.

The overall objective of the study, therefore, was to develop and test a comprehensive methodology to assess and evaluate the performance of existing RWH in arid and semi-arid regions. To achieve this goal, we developed a new RWH evaluation and decision support tool. In this tool, engineering, biophysical and socio-economic criteria were taken into account to assess the performance of existing RWH, using the Analytical Hierarchy Process supported by GIS. To develop and test this assessment tool, the Oum Zessar watershed in south-eastern Tunisia was selected as a case study. Jessour and Tabias are the most common RWH techniques in the Oum Zessar watershed and they are used in our methodology.

2. Materials and Methods

2.1. Case Study: Wadi Oum Zessar

To test the RWH evaluation tool we conducted a case study in the Wadi Oum Zessar watershed located in Medenine province in the south-eastern part of Tunisia (Figure 1). The Wadi Oum Zessar watershed has a surface area of 367 km^2. The area is characterized by a low arid Mediterranean climate, with an average annual rainfall of 150–230 mm·y^{-1}, and average annual temperature of 19–22 °C.

Rainfall occurs mainly in winter (40%), autumn (32%) and spring (26%), while summer is almost rainless [22].

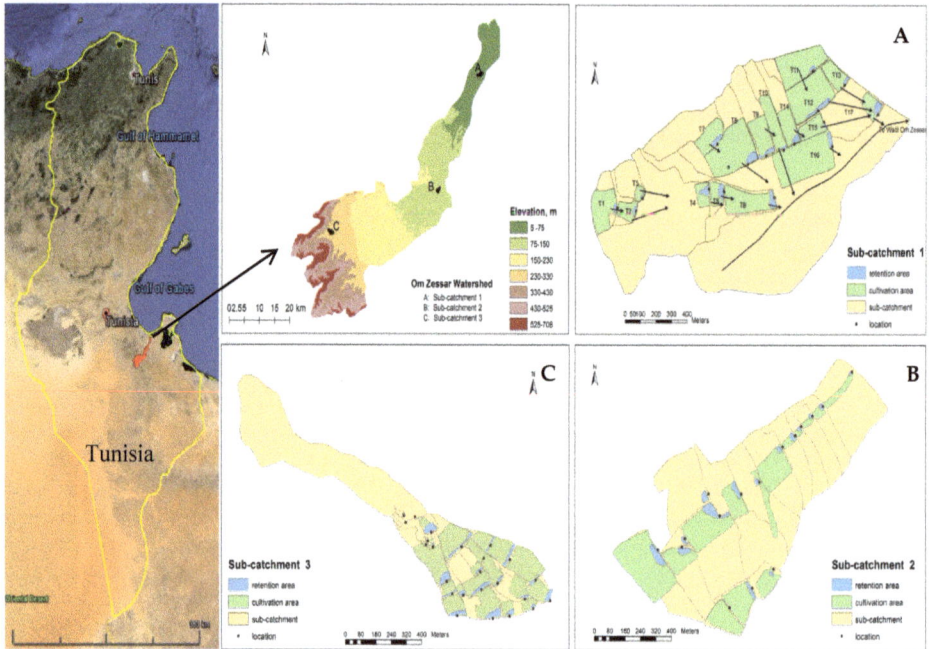

Figure 1. Location of Oum Zessar and test sub-catchments; (**A**) Sub-catchment 1; (**B**) Sub-catchment 2, and (**C**) Sub-catchment 3.

Several types of RWH exist in the study area to satisfy water requirements for agriculture and ground water recharge. The most common RWH systems in the region are jessour and tabias; spreading of flood water and groundwater recharge structures in the wadi beds are applied too [23].

To test the RWH evaluation tool, three representative sub-catchments were selected based on four criteria.

i   Representative of the geographic distribution of our watershed; one located in upstream another in the midstream and one in downstream.

ii   Representative of the different types (jessour and tabias), scale (small and large) and age of RWH systems (new and old).

iii Source and destination of collected rainwater for each sub-catchment.

iv Accessibility; easy to access physically and acceptance of the local people.

These three sub-catchments are located in the downstream (Sub-catchment 1), middle (Sub-catchment 2), and upstream (Sub-catchment 3) of Oum Zessar watershed as shown in Figure 1. Each jessr (singular of jessour) or tabia consists of three parts: the impluvium or catchment area providing the runoff water; the terrace or cultivation area where the runoff water is collected and crops or trees are grown; and the dyke, which is a barrier to catch water and sediment. Each dyke has a spillway (*menfes* if the spillway is located on one or both sides and *masref* if the spillway is located in the middle of the dyke) to regulate water flow between dykes (see Figure 2).

Figure 2. (**A**) An example of jessour (Ouessar 2007) and (**B**) properties of jessr.

2.2. General Description of the RWH Evaluation Decision Support Tool

This research aims to develop a more comprehensive and relevant evaluation tool for RWH structures. To achieve this goal, we developed a simple and robust assessment tool for the evaluation of RWH sites (structures) which is inexpensive, simple to apply, reliable and flexible with different criteria and easy to adapt to various RWH techniques and regions. The Analytical Hierarchy Process (AHP) forms the base for this tool.

The AHP is a multi-criteria decision making method, providing a structured technique for organizing and analyzing complex decisions, based on mathematics and expert knowledge [24]. It was developed by Thomas Saaty in the 1970s and, since then, has been applied extensively in different disciplines. The main principle of AHP is representing the elements of any problem hierarchically to show the relationships between each level. The uppermost level is the main goal (objective) for resolving a problem and the lower levels are made up of the most important criteria that are related to the main objective. Pairwise comparison matrixes are constructed and scaled in preference from 1 to 9 for each level. Then, the consistency of each matrix is checked through the calculation of a consistency ratio (cr). The cr should be smaller or equal to 10% [25]. The weight for each criterion and the cr are determined, then all matrixes are solved.

2.3. Methodology Overview

AHP is particularly useful in multi-index evaluation and consists in our RWH evaluation tool of the following steps:

i   Describe the main objective of the intervention;
ii   Identify the biophysical, engineering (technical) and socio-economical main and sub-criteria;
iii   Develop a decision hierarchy structure;
iv   Collect and process the data for each sub-criteria;
v   Classify the values for each sub-criteria in terms of suitability classes;
vi   Apply the pairwise comparison matrix to identify priorities (weights) for each criterion;
vii   Calculate the RWH performance (suitability);
viii   Check the results with the stakeholders; and
ix   Decide based on conclusions and recommendations

2.3.1. Description of the Main Objective of the Intervention

In our case study, the main objective is to collect and store runoff water during the rainy season to enable farmers to grow profitable crops and mitigate drought spells in arid and semi-arid regions.

2.3.2. Identification of the Main and Sub-Criteria

This step formulates the set of criteria for the assessment based on the main objective. All major aspects should be represented, but the set should be as small as possible (simple and flexible). In addition to engineering (technical) aspects, social and economic aspects should also be included. Furthermore, the set of criteria has to be operational (e.g., measurable) and not redundant (the set should not count an aspect more than once).

In this study, we looked for criteria that represent the key parameters affecting the performance of RWH interventions and which could be applied to different sites and techniques. The parameters we were concerned with were based on the general definition of RWH, *i.e.*, a method for inducing, collecting, storing and conserving local surface runoff for agriculture in arid and semi-arid regions [2], and information found in literature studies. The main selected criteria and sub-criteria are shown in Figure 3, and reflect the following questions:

i How suitable is the local climate for RWH (Climate and drainage)?
ii What is the engineering (technical) performance of the RWH intervention (Structure design)?
iii How suitable is the location for RWH (Site characteristics)?
iv How well does the RWH satisfy the water demand (Reliability)? and
v How well does the RWH technique fit in with the social economic context (Socio-economic criteria)?

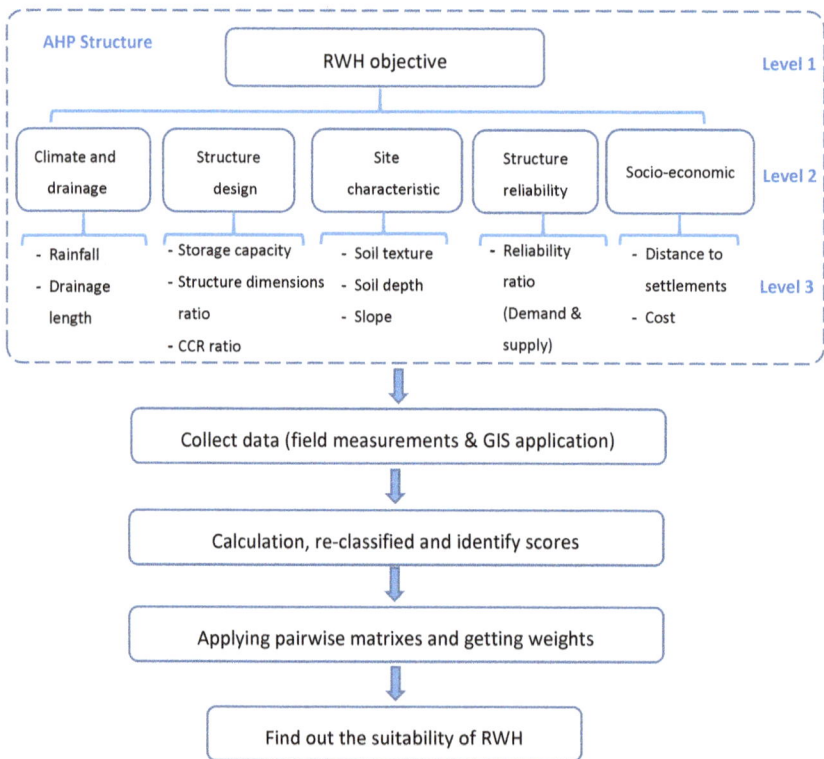

Figure 3. The schematic of the RWH (Rain Water Harvesting) suitability model, criteria and hierarchy structure for two methodologies. Method 1 consists of three levels and method two consists of two levels (Level 1 and Level 3).

Sub-criteria were chosen based on the relation with the main criteria (above), field investigations, expert discussions and literature studies.

2.3.3. Development of the Decision Hierarchy Structure

In this step, the main criteria and sub-criteria are arranged in a multilevel hierarchical decision structure. In this study case, the objective of the RWH (jessour and tabias) represents the first level. The second level contains the main criteria for the assessment. These criteria define the aspects by which the intervention is assessed e.g., how it fits within the local conditions (climate, drainage length and landscape), functionality and reliability based on the engineering design, and socio-economic aspects. The sub-criteria used to measure the performance of each main criterion are represented in the third level. Figure 3 shows the structure of the applied methodology for our case study.

2.3.4. Collection and Processing of the Data for Each Sub-Criteria

The definition, data collection, field measurements, storage and processing of data, as well as the calculations used for each criterion is explained in detail in the Section 2.4.

2.3.5. Classification of the Values for Each Sub-Criteria in Terms of Suitability Classes

Due to the variety of measurements and scales for the different criteria, a comparable scale between criteria must be identified before applying AHP tools. For instance, rainfall depth is measured in mm while soil texture is measured by the percentage of clay content. Therefore, the selected criteria were re-classified into five suitability classes, namely, 5 (very high suitability), 4 (high suitability), 3 (medium suitability), 2 (low suitability), and 1 (very low suitability). For example, suitability Class 3 is considered to be acceptable performance, while suitability Class 1 means that the RWH does not work well and that one or all criteria that caused this insufficient performance need improvement. Table 1 shows the scores assigned based on discussions and consultations with experienced people and information found in the literature.

2.3.6. Application of Pairwise Comparison Matrix to Identify Priorities (Weights) for Each Criteria

After assignment of scores, the weight for each criterion was determined by applying AHP with the pairwise comparison matrix. Pairwise comparison concerns the relative importance of two criteria involved in determining the suitability for a given objective. A pairwise matrix is first made for the main decision criteria being used. Other pairwise matrixes are created for additional criteria levels. The comparison and rating between two criteria are conducted using a 9-point continuous scale, the odd values 1, 3, 5, 7, and 9 correspond respectively to equally, moderately, strongly, very strongly and extremely important criteria when compared to each other. The even values 2, 4, 6 and 8 are intermediate values [26]. During pairwise comparison, criteria were rated based on the literature review, information from the field survey and discussions with stakeholders and experts. The final weight calculation requires the computation of the principal eigenvector of the pairwise comparison matrix to produce a best-fit set of weights. The consistency of each matrix, which shows the degree of consistency that has been achieved by comparing the criteria, was checked through the calculation of consistency ratio (cr). The cr should be smaller or equal to 10%, otherwise they are judged as not consistent enough to generate weights and, therefore, have to be revised and improved [25].

Table 1. Classification, suitability levels and scores for each criterion for assessment of existing RWH sites in arid and semi-arid regions. Each value, class and score were rated based on the literature review, information from the field survey and discussions with stakeholders and experts.

Criteria (Indicator)	Classes	Values	Scores Jessr/Tabia	Scores (Tabia) *
Rainfall (mm·y⁻¹), more rainfall on any particular area means higher possibilities of harvesting part of it. [27]	Very low suitability	<100	1	
	Low suitability	100–175	2	
	Medium suitability	175–250	3	
	High suitability	250–325	4	
	Very high suitability	>325	5	
Drainage length (m), the distances from the water courses to each dyke (short distance means fewer losses). [15]	Very high suitability	0–50	5	
	High suitability	50–125	4	
	Medium suitability	125–200	3	
	Low suitability	200–300	2	
	Very low suitability	>300	1	
Storage capacity ratio (-), the ratio between the total volume of water inflow and existing storage capacity. The ratio that is close to one is ranked as highly suitable.	Over requirement (too large a storage capacity area)	<0.5	2	
	Sufficient	0.5–1.0	4	
	Optimum requirement	1.0–2.0	5	
	Critical	2.0–4.0	3	
	Very critical requirement (too small a storage capacity area)	>4.0	1	
Structure dimensions ratio (-), the ratio between the required design height and the existing height of dykes or barriers for each RWH structure. The ratios that are close to one are ranked as highly suitable	Over design (existing height is double what is required)	<0.5	3	
	Suitable	0.5–0.75	4	
	Optimum	0.75–1.0	5	
	Under design	1.1–1.25	2	
	Critical (existing height is lower than required)	>1.25	1	
Catchment to cropping area (CCR ratio (-))	Medium suitability	<0.5	2	
	Very high suitability	0.5–0.75	4	
	Suitable	0.75–1.25	5	
	Low suitability	1.25–2.0	3	
	Very low suitability	>2.0	1	
Soil texture (Clay content %) [28]	Very high suitability (Clay)	>20	5	
	High suitability (Silty clay)	15–20	4	
	Medium suitability (Sandy clay)	11–15	3	
	Low suitability (Sandy clay loam & sandy loam)	8–11	2	
	Very low suitability (other)	<8	1	

* Different suitability classes for slopes between jessour and tabias.

Table 1. Cont.

Criteria (Indicator)	Classes	Values	Scores Jessr/Tabia	Scores (Tabia) *
Soil depth(m) [1]	Very deep	>1.5	5	
	Deep	0.9–1.5	4	
	Moderately deep	0.5–0.9	3	
	Shallow	0.25–0.5	2	
	Very shallow	<0.25	1	
Slope (%) [29]	Flat	<1.5	1	2 *
	Undulating	1.5–3	3	5
	Rolling	3–5	4	4
	Hilly	5–10	5	3
	Mountainous	>10	2	1
Reliability ratio (-), the ratio between the total demand and the total supply of water. High suitability scores for the ratio are close to one	Sufficient (required water is largely less than supply)	<0.35	2	
	Medium Sufficient	0.35–0.75	4	
	High Sufficient	0.75–1.1	5	
	Large deficit	1.1–1.75	3	
	Very large deficit (required water is largely higher than supply)	>1.75	1	
Distance to settlements (km), highest scorers are ranked to the closest distance to the settlements (high suitability). [6]	Very high suitability (too short a distance)	<0.5	5	
	High suitability	0.5–0.75	4	
	Medium suitability	0.75–1.25	3	
	Low suitability Very low suitability (too far a distance)	1.25–1.75	2	
		>1.75	1	
Cost ($·m⁻³ of water), low cost indicates high scores (profitable). Costs are estimated based on the WOCAT database [30] and farmer interviews	Very high cost (very low suitability)	>12	1	
	High cost	9–12	2	
	Medium cost	6–9	3	
	Suitable cost	3–6	4	
	Profitable cost (very high suitability)	<3	5	

* Different suitability classes for slopes between jessour and tabias.

To find out the final weight for each criterion and the cr, we solved the pairwise matrixes mathematically. The results of the main criteria from the pairwise comparison and the final weight are presented in the results section.

In this study, two methods were applied. In the first, the hierarchy structure consists of all three levels; the objective, main criteria (5 criteria) and sub-criteria (11 criteria). In the second method, the hierarchy structure consists of just two levels: the objective and the sub-criteria (11 criteria). By applying these two methods, the understanding of the relation between each criterion and its reflection on the main objective becomes much clearer, and they confirm the flexibility of AHP to adopt different criteria on multi-levels. Moreover, this will give an insight into whether there are any mistakes and how they will be distributed or fixed, and gives more reliability and confidence in our methodology for adoption in different regions and/or for different criteria.

2.3.7. Calculation of the RWH Performance (Suitability)

The next step in the assessment methodology is the calculation of the overall suitability for each RWH site. The overall RWH suitability was calculated by applying the following formula:

$$S = \sum_{i=1}^{n} W_i \, X_i \tag{1}$$

where: *S*: suitability; *W*: weight of criteria *i*; *X*: score of criteria *i*; *i*, *n*: number of criteria

The overall suitability will be classified also from 1 to 5, namely, 5 (very high suitability), 4 (high suitability), 3 (medium suitability), 2 (low suitability) and 1 (very low suitability).

2.3.8. Discussion of the Results with Stakeholders

It is important to check the results with the stakeholders, including the preliminary conclusions and recommendations. If felt that something is missing or has changed, additional measurements or recalculation with different weights might be necessary. Thereafter, results have been presented again to the local stakeholders for discussion and approval.

2.3.9. Decision Making Based on Conclusions and Recommendations

The main results of the assessment will give insight into if and how a RWH structure can be improved to increase its performance. Once there is general agreement on the results between stakeholders and scientists, a well-founded decision can be made on what structure needs to be improved for better performance of the RWH system.

2.4. Data Collection

Different data sources were used. Meteorological as well as other biophysical data, was collected from the Institute des Régions Arides (IRA) in Tunisia. Field measurements were carried out in the Wadi Oum Zessar during the period from December 2013 through March 2014. An open structure interview was made with key stakeholders (41 landowners and farmers) and discussions with people working and having experience with RWH (15 experts), particularly the engineers from the Regional Department of Agriculture in Medenine. A pairwise matrix was established and the relative weights for each criterion and suitability rank for classes are assigned as shown in Table 1. GIS was also applied to extract data that are needed in our methodology. All collected and measured data were stored and processed using Excel software.

2.4.1. Climate and Drainage Data

Rainfall

Rainfall is one of the major components in any RWH system, with the magnitude of rainfall playing a significant role in assessing the RWH suitability for a given area. In arid and semi-arid regions, rainfall varies greatly in time and space. RWH systems can only function if there is sufficient rainfall in the catchment area to be stored somehow. Average monthly rainfall for the period 1979–2004 was collected from IRA for 7 meteorological stations in the Wadi Oum Zessar watershed, namely Ben Khedache, Toujan Edkhile, Allamat, Koutine, Sidi Makhlouf, Ksar Hallouf and Ksar Jedid. The rainfall amount in the three test sub-catchments was determined by applying the Inverse Distance Weight (IDW) function from ArcGIS 10.0 to interpolate the data from these stations. The rainfall depth data was then reclassified and scored as shown in Table 1. Areas with high annual rainfall are ranked as highly suitable.

Drainage Length

Since RWH interventions (especially jessour and tabias) are located on the hydrographic network and their location is influenced by topography, the distance from the water course has a significant role in the assessment of RWH performance. In this study, the distance from a RWH site to the drainage networks is used to represent the runoff suitability. By determining the location of the furthest point contributing to runoff [31], the drainage system was classified to each of the RWH sites (short distance means fewer water losses). The distances from the water courses to each dyke were measured using Google earth image and ArcGIS software.

2.4.2. Structure Design

Storage Capacity

One of the main principles of RWH is storing water to mitigate drought effects in dry seasons. Technically, the volume of water harvested and the amount retained over a reasonable duration of time is one indicator of the performance of RWH.

Potential runoff (V_1 in m^3) from a catchment area was calculated by:

$$V_1 = 0.001 \times C \times P \times A \tag{2}$$

C: The mean annual runoff coefficient (-); equal (0.18) based on the simulations done by Schiettecatte *et al.* [32].

P: The mean annual precipitation (mm)

A: The catchment area (m^2)

The total volume of water inflow (V_i) is, therefore:

$$V_i = V_1 + V_2 + V_3 \tag{3}$$

where V_2 (m^3) is the overflow from upstream dyke(s) and V_3 (m^3) is the volume of rainfall onto the storage area.

During the field measurements, the retention area and maximum potential depth of water (height of spillway) were measured with GPS and measuring tape. Then, the existing storage volumes were calculated (by multiplying the retention area by spillway height). Finally, the ratio between the total volume of water inflow (V_i) and existing storage capacity were calculated and scored. If the ratio, for example, equals 1–2, it means that the total inflow volume will be similar to the storage capacity or there is excess water that will be an overflow to the downstream. Therefore, the ratios that are close to one are ranked as highly suitable (Table 1).

Structure Dimensions

The dimensions of RWH structures are very important for achieving stability, controlling flood hazard and water supply. Furthermore, the primary goal of a structure is to harvest water for irrigation crops; the secondary goal is for flood protection. In this study, we assessed the existing height of dykes or barriers for each RWH structure and then compared this with the theoretical (required) design height.

The existing dyke's height for each site was measured in the field. The total volume of water that could be collected behind each dyke was calculated as noted in the previous section. The effective dyke height was calculated using this information. The free board, the vertical distance between the top of the dam and the full supply level, was calculated using standard dam design principles and added to the effective dyke height to determine the theoretical design height for each site. The ratio between existing and design dyke height was calculated and scored, as shown in Table 1.

Catchment to Cropping Area

To provide sufficient water to the crops, the terrace area should be not too large and the impluvium area should be enough. Therefore, an optimal ratio between impluvium area and terrace area has to be found. Depending on effective rainfall and runoff rates, the ratio between the catchment (impluvium) and cropping (terrace) area (*Ca/C*) can be determined. According to Schiettecatte *et al.* [32], the minimum ratio (*Ca/C*) "impluvium area/terrace area" (design) can be calculated by:

$$\frac{Ca}{C} = (WR - P)/CP \tag{4}$$

where *WR* is the annual crop water requirement, *P* is the average annual precipitation (mm) for the period 1979–2004, and *C* is the average annual runoff coefficient (0.18) of dry soil and wet soil which was measured by Schiettecatte *et al.* [32]. For olive trees, the *WR* is estimated to be 500 $mm \cdot y^{-1}$ [32]. Catchment area (impluvium) and cropping area were delineated with GPS in the field, and the areas were calculated using ArcGIS. At the end, the CCR ratio between the design and existing "impluvium area/terrace area" were calculated and scored.

2.4.3. Site Characteristics

Soil Texture

Soil texture is a very important factor in selecting, designing and assessing the performance of RWH. Soil texture affects both the infiltration rate and surface runoff. The textural class of a soil is determined by the percentage of sand, silt and clay. Soil texture also determines the rate at which water drains through a saturated soil; for instance, water moves more freely through sandy soils than it does through clayey soils. High infiltration rates such as with sandy soil are not suitable for RWH structure. Clay soils have a greater water holding capacity than sandy soils, therefore, soil with high water holding capacity are more suitable for RWH. Indeed, Mbilinyi *et al.* [33] and others conclude that clay soil is best for water storage due to its low permeability and ability to hold the harvested water.

In this research, the terrace area was sampled at different sites (based on the size of terrace area, 1–3 samples for each site) and at depths of up to 1.3 m. The samples were taken to the IRA laboratory and analyzed. The clay contents (%) were measured, rated and classified into five suitability classes, as shown in Table 1.

Soil Depth

Soil should be deep enough to allow excavation to the prescribed depth for RWH, to ensure both adequate rooting development and storage of the harvested water. Critchley and Siegert [20] and Kahinda *et al.* [1] used soil depth as one criterion for selecting potential sites for RWH. Both soil depth and soil texture determine the total soil water storage capacity, which controls the availability of water for crops during the dry periods [9]. We measured soil depth in the field using a steel bar hammered into the ground until it could go no further and by checking the soil levels between two successive terraces. Then, soil depth data were categorized and classified into five suitability classes, as shown in Table 1.

Slope

Slope is also a major factor in site selection, implementation and assessment of RWH. It plays a significant role in runoff and sedimentation quantity, the speed of water flow and quantity of material required to construct the dyke structure (dyke's height).

Using DEM (30 m resolution) and ArcGIS 10.0, the slope was extracted for each catchment area and reclassified. Due to the large variety of slope values between jessour and tabias, different suitability classes were used for each type as shown in Table 1.

2.4.4. Structure Reliability

The relation between the demand and supply of water (reliability) is a good indicator of the performance of a RWH structure. Based on the function (purpose) of each technique, the demand for each RWH site was calculated. In our case, the main purpose of RWH is for on-site crop production.

The total demand was calculated by estimating the crop water requirements (evapotranspiration ET_c) plus losses to downward percolation, based on the field measurements by Schiettecatte *et al.* [32] in the same watershed.

$$The\ total\ demand = ET_c + Downward\ percolation \tag{5}$$

Schiettecatte *et al.* [32] applied the Penman-Monteith method to calculate potential evapotranspiration (PET) and used data from the meteorological station at Medenine to calculate the average PET values over the period 1985–1995.

The maximum crop evapotranspiration (ET_c) was calculated by:

$$ETc = PET \times k_c \tag{6}$$

where k_c is the crop coefficient. Table 2 above shows the values for PET, ET_c and k_c.

Table 2. Rainfall, potential evapotranspiration (PET), maximum crop evapotranspiration (ET_c) and olive crop coefficient k_c results [32], by applying the Penman-Monteith method and using meteorological data from Medenine station.

Month	Rainfall (mm)/year	PET (mm)	ET$_c$ (mm)	k_c for Olive
January	37.5	69.6	27.8	0.40
February	30.6	88.6	35.4	0.40
March	40.0	121.2	66.7	0.55
April	16.3	159.3	79.6	0.50
May	11.2	198.4	89.3	0.45
June	1.0	213.5	85.4	0.40
July	0.0	234.8	82.2	0.35
August	2.0	220.9	77.3	0.35
September	17.1	166.6	75.0	0.45
October	23.0	126.8	63.4	0.50
November	19.9	91.1	41.0	0.45
December	36.7	67.4	26.9	0.40

The infiltration ratios were used to calculate the downward percolation based on the soil texture results, as shown in Table 3 [34].

Table 3. Typical values for final infiltration rate for various soil textures [34].

Soil Type	Infiltration Rate (mm·h^{-1})
Coarse sand	>22
Fine sand	>15
Fine sandy loam	12
Silt loam	10
Silty clay loam	9
Clay loam	7.5
Silty clay	5
Clayey soil	4

From the relation between storage capacity and total runoff volume from Equation (2), the total potential volume of supply water was calculated. Reliability was calculated as the ratio between total demand and the total supply of water for each site.

2.4.5. Socio-Economic Criteria

The success of an intervention depends not only on technical aspects but also on how well it fits within the stakeholder's social context and the economic benefit it provides him/her. Bamne [35], Al-Adamat [27] and Nasr [36] argued that one of the main reasons we do not use RWH sufficiently in the Middle East and North Africa is insufficient knowledge of the socio-economic contexts. There are several socio-economic criteria such as ownership, family size, education *etc.*, and to identify good indicators for socio-economic conditions in relation to the functioning of these RWH systems is much more difficult than the biophysical ones. In this case study based on the literature studies and expert discussion, we are using distance to the settlements and cost per cubic meter of water as the socio-economic criteria influencing how suitable the intervention is for the main stakeholders.

Distance to Settlements

Since the local community is targeted in this study, the distance to the settlements is an important parameter in the design, selection and assessment of the RWH suitability [6]. We assumed that the distance to their home would influence the way they manage this system. Each farmer has scattered farming fields at a radius of about 20 m–1 km from his house. Therefore, it is very logical that the closer the field, the easier are the maintenance operations, particularly in the mountain zones where transportation is difficult. The distance for each site was measured using the image from Google earth and the ArcGIS program. Thereafter, as with other criteria, the values were reclassified and scored.

Cost Per Cubic Meter of Water

Cost plays a significant role in the design and assessment of RWH sites. In order to assess the cost effectiveness of each structure, the establishment and annual maintenance costs for each site were calculated. The actual costs for each structure were not available; the main problem with the jessour and tabia is that they do not have fixed designs (different shapes and sizes). Therefore, it is difficult to calculate the exact cost for each structure. Thus, the costs have been estimated by using the best available resources. The cost for each jessr or tabia was calculated based on the World Overview of Conservation Approaches and Technologies (WOCAT) database [30] and interviews with the local farmers. The costs for each jessr/tabia include the establishment and maintenance cost per year. The establishment costs consist of dyke construction, plantations, spillway construction for jessour and diversion channels and terracing for tabia. The maintenance costs consist of crop and tree maintenance, dyke and spillway maintenance, repairs and reconstruction. The overall costs for jessre per year are 3000 US$ for establishment and 900 US$ for maintenance. Whereas, 670 and 200 US$ for establishment and maintenence for tabia per year, respectively [30]. Based on the field measurements, the length for each jessr/tabia was measured and then the cost for each meter length of jessr/tabia was estimated.

These costs are similar to the values that were discussed with local farmers. The volume of collected water in each storage area and maintenance and construction costs of the jessour/tabias were used to calculate the cost per cubic meter of water, which was then classified and scored.

2.5. Application of the Assessment Tool for Different Test Sub-Catchments

We first tested our methodology on a catchment that has only one type of RWH structure. Sub-catchment one has just 17 tabias and no jessour and a total area of about 20 ha. It is located in the downstream area of the Oum Zessar watershed, as shown in Figure 1.

To further validate the methodology and criteria, we applied it on the other two sub-catchments, which have different characteristics. The second sub-catchment is located in the middle of Wadi Oum Zessar and has 16 RWH structures, 9 tabias followed (downstream) by 7 jessour, and a total area of about 19 ha. Sub-catchment three is located in the upstream part of Wadi Oum Zessar, with 25 RWH—8 jessour followed by 17 tabias—and a total area of about 45 ha.

3. Results

All the collected data for each site were stored and analyzed in Excel. The results for each criterion were then classified according to the five classes as defined in Table 1. Figure 4 shows the scores percentages (5 scores) of each sub-criteria (11 criteria) for all 58 sites. The rainfall criterion got a score 3 in all sites since there was no big difference in rainfall pattern nor amount (175–185 mm·y^{-1}) in the three sub-catchments due to the relatively small area. The criteria related to the design structure, like dimensions, storage capacity, CCR, drainage flow and costs got a high percentage of scores of 1 in many sites. More details about suitability and scores for the three sub-catchments are explained in the following sections.

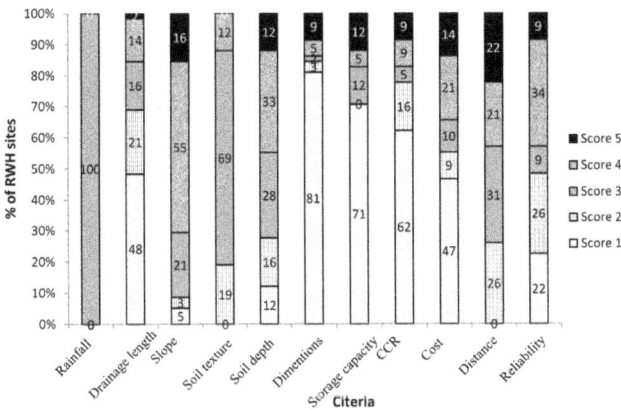

Figure 4. The score percentages for each criterion in all RWH sites (*n* = 58), the five scores were determined based on classifications by experts and previous studies.

3.1. AHP & Suitability

During pairwise comparison, criteria were rated based on the literature review, interviews with key stakeholders, field survey information and discussions with people working and having experience with RWH, as shown in Table 4. For instance, the reliability and socio-economic criteria have similar relative importance to the main objective of the RWH system, as shown in this Table, and each of them has 1 as a relative importance rate.

Table 4. The pairwise comparison matrix for the main criteria (Method 1).

	Climate and Drainage	Structure Design	Site Characteristics	Reliability	Socio-Economic
Climate and drainage	1	2	1	3	2
Structure design	1/2	1	1	1	2
Site characteristics	1	1	1	2	3
Reliability	1/3	1	1/2	1	1
Socio-economic	1/2	1/2	1/3	1	1

A pairwise matrix was established and the relative weights for each criterion and suitability rank for classes are assigned as shown in Figure 5 and Table 1. The climate and rainfall criteria received the highest weights in both methods (three levels and two levels AHP). The values for each criterion were calculated and reclassified based on the 5 suitability classes and Equation (1) was applied to get the final suitability score for each site.

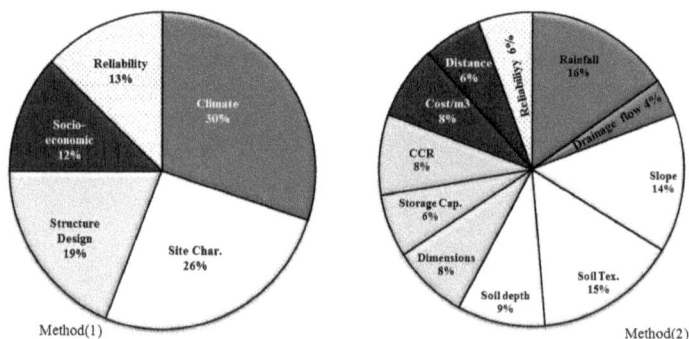

Figure 5. The weights for main criteria in two methods: Method 1 consists of three levels, the objective in the first level, five main criteria in the second level and 11 sub-main criteria in the third level; while Method 2 has just two levels, the objective in the first level and the 11 indicators (main criteria) on the second level.

3.2. Test Results Sub-Catchment 1

Table 5 shows measurements and scores for each criterion for the tabias receiving the highest (9 and 14) and lowest (10 and 15) suitability scores when AHP Method 1 was applied (before applying Equation (1)).

Figure 6A shows the overall suitability scores and the suitability score for each criterion based on Method 1 (three levels) after applying Equation (1). The highest overall score was 3.32 (medium suitability) for tabia 9, whereas the lowest score was 2.04 (low suitability) in tabia 10.

Design criteria (structure dimensions, storage capacity and catchment area to cropping area) are playing a significant (negative) role in the overall RWH suitability for most of the tabias in Sub-catchment 1. These sites scored the lowest on design criteria, resulting in the low overall performance of these RWH sites. This result confirmed the observations of performance in the field.

A possible reason for the poor design is a lack of selection procedure for suitable RWH sites in combination, in this case, with structures built without a proper engineering design. Figure 6B shows the suitability scores for each criterion without multiplying by the weights.

Table 5. The measurements and scores for each criterion (indicator) for the tabias receiving the highest (9 and 14) and lowest (10 and 15) suitability scores in Sub-catchment 1, when AHP Method 1 was applied (before applying Equation (1)).

Criteria (Indicator)	Sub-Catchment 1, Tabia No.							
	High				Low			
	9		14		10		15	
	M *	S **	M	S	M	S	M	S
Rainfall (mm·y^{-1})	180.00	3	180.00	3	180.00	3	180.00	3
Drainage length (m)	255.00	2	243.00	2	257.00	2	340.00	1
Slope (%)	3.50	4	7.90	3	5.76	3	4.60	4
Soil Texture (clay contents %)	14.30	3	12.60	3	8.70	2	11.10	3
Soil depth (m)	0.80	3	0.95	4	0.80	3	0.75	3
Structure dimensions ratio (-)	0.93	5	1.03	5	4.88	1	4.30	1
Storage Capacity ratio (-)	2.49	3	3.02	3	34.00	1	34.50	1
CCR ratio (-)	3.80	1	4.20	1	1.30	3	9.60	1
Cost ($·m^{-3} of water)	5.90	4	6.40	3	48.00	1	43.00	1
Distance to settlements (km)	1.20	3	1.24	3	1.56	2	1.32	2
Reliability ratio (-)	0.50	4	0.68	4	4.46	1	2.47	1

* measurements/calculation data; ** scores.

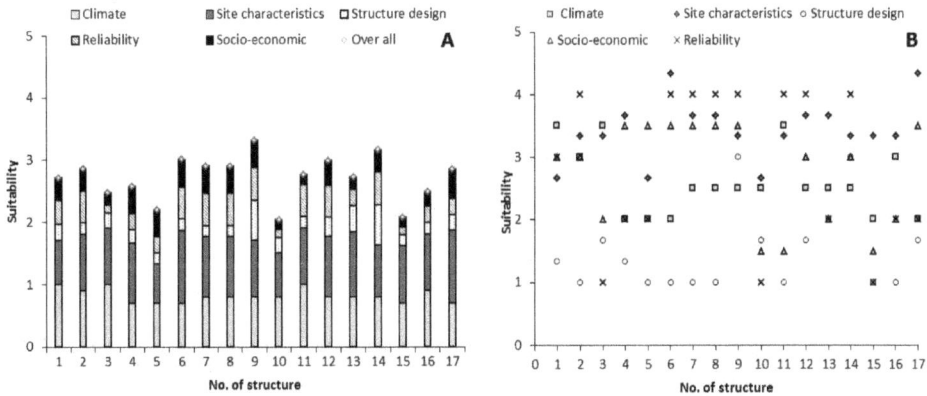

Figure 6. The overall suitability and the suitability for each criterion in each site of Sub-catchment 1 (Method 1), the left figure (**A**) shows the results after applying weights and Equation (1), the right figure (**B**) shows the scores without applying weights to compare weight effecting on the suitability scores for each criteria as shown in the left figure.

In Method two (two levels), the pairwise matrix was applied directly on the sub-criteria. Table 6 shows the overall suitability and the suitability for each criterion for the highest (9 and 14) and lowest (10 and 15) scoring tabias using this method. Once again, the design criteria of dimension and storage capacity had a significant negative impact on the difference between the high-scoring and low-scoring tabias. However with Method 2, CCR did not stand out as a differentiating factor, but reliability and cost did.

Table 6. The overall suitability and the suitability for each criterion for the highest (9 and 14) and lowest (10 and 15) scoring tabias in Sub-catchment 1, according to Method 2 and after applying Equation (1).

Criteria	Sub-Catchment 1, Tabia No.			
	High		Low	
	9	14	10	15
Rainfall (mm·y^{-1})	0.465	0.465	0.465	0.465
Drainage length (m)	0.076	0.076	0.076	0.038
Slope (%)	0.572	0.429	0.429	0.572
Soil Texture (clay contents %)	0.450	0.450	0.300	0.450
Soil depth (m)	0.279	0.372	0.279	0.279
Structure dimensions ratio (-)	0.395	0.395	0.079	0.079
Storage Capacity ratio (-)	0.195	0.195	0.065	0.065
CCR ratio (-)	0.083	0.083	0.249	0.083
Cost ($·m^{-3} of water)	0.300	0.225	0.075	0.075
Distance to settlements(km)	0.186	0.186	0.124	0.124
Reliability ratio (-)	0.228	0.228	0.057	0.057
Overall score	3.23	3.10	2.20	2.29

3.3. Test Results Sub-Catchments 2 and 3

The suitability scores for each criterion and overall from applying Method 1 (three levels) in Sub-catchments 2 and 3 are shown in Figure 7. The socio-economic criteria played a significant role in the assessment methodology here, especially for jessour in these sub-catchments (8–16 in Sub-catchment 2 and 1–8 in Sub-catchment 3) because of the high cost of implementing and maintaining the RWH compared with the relatively small area and low quantity of water retained behind the dykes. Moreover, these techniques are most common in this region especially in the mountain areas. They seem to be the most suitable techniques to mitigate flood hazard, additionally, the stakeholders consider them to be part of their heritage.

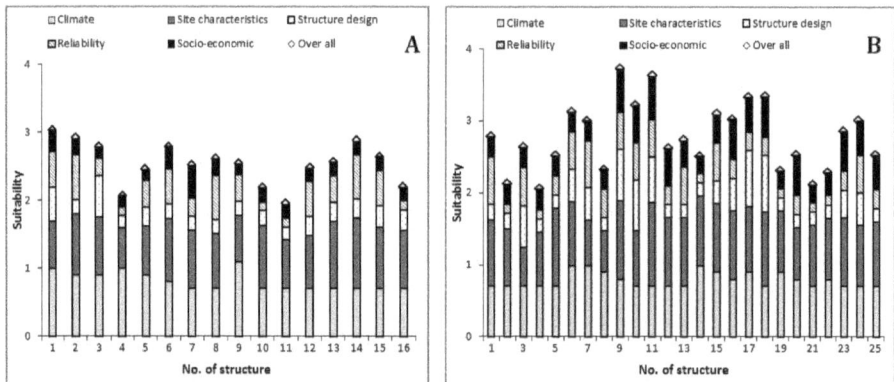

Figure 7. The overall suitability and the suitability for each criterion in each site in Sub-catchment 2 (**A**) and 3 (**B**) according to Method 1. In Sub-catchment 2 the overall suitability is hovering between 1.94 and 3.03 and site suitability in most of the sites got the highest scores among other criteria, and in Sub-catchment 3 the overall suitability is hovering between 2.05 and3.72 and the site suitability almost got the highest scores too.

Table 7 shows the individual criteria and overall suitability scores for the highest and lowest scoring sites in Sub-catchments 2 and 3 after applying method 2.

Table 7. The individual criteria and overall suitability scores for the highest and lowest scoring sites in Sub-catchments 2 and 3 after applying Method 2.

Criteria	Tabia/Jessr No.			
	Sub-Catchment 2		Sub-Catchment 3	
	High	Low	High	Low
	14	11	11	21
Rainfall (mm·y^{-1})	0.465	0.465	0.465	0.465
Drainage length (m)	0.038	0.038	0.038	0.038
Slope (%)	0.572	0.429	0.715	0.572
Soil Texture (clay contents %)	0.600	0.450	0.600	0.450
Soil depth (m)	0.372	0.186	0.372	0.186
Structure dimensions ratio (-)	0.079	0.079	0.316	0.079
Storage Capacity ratio (-)	0.065	0.065	0.260	0.065
CCR ratio (-)	0.332	0.083	0.083	0.083
Cost ($·m^{-3} of water)	0.075	0.075	0.375	0.075
Distance to settlements (km)	0.186	0.186	0.310	0.248
Reliability ratio (-)	0.285	0.057	0.228	0.057
Overall suitability	3.07	1.92	3.76	2.32

Catchment to cropping areas ratio (CCR) has a significant effect on overall suitability scores in Sub-catchment 2, whereas in Sub-catchment 3 there was not a difference in CCR between the high and low scoring structures. Moreover, slope played an important role in the overall scores in Sub-catchment 3 but not in Sub-catchment 2 (Table 7).

3.4. Comparison of Methods 1 and 2

A comparison between the two methods of applying AHP (three and two levels structure) in our methodology is shown in Figure 8. Although the results are very similar, Method 2 gives a slightly higher score for the jessour in Sub-catchment 2 (jessour 10–16) and Sub-catchment 3 (jessour 1–8).

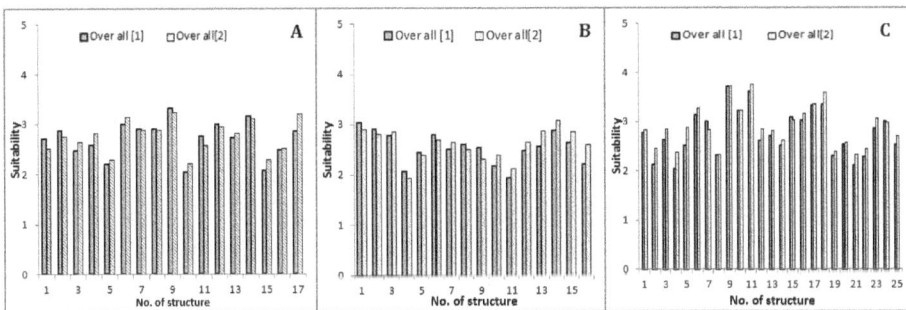

Figure 8. The comparison between overall scores for the two methods in the three test sub-catchments (**A**) Sub-catchment 1, (**B**) Sub-catchment 2 and (**C**) Sub-catchment 3. The results are very similar, Method 2 gives a slightly higher score for the jessour/tabia in Sub-catchment 1 (tabias 3, 4, 6 and 17), Sub-catchment 2 (tabia 3 and 7, jessour 10–16) and Sub-catchment 3 (jessour 1-6 and tabias 11, 18 and 23).

The consistency of each matrix was calculated using the consistency ratio (cr). For the main criteria matrix in Method 1 cr was 2.9% and for the second method cr was 2.4%.

The principles of AHP call for the cr to be smaller or equal to 10%, therefore the cr values were acceptable.

These results suggest that both methods are good and easy to adapt to different criteria, thus researchers can apply either of the two methods.

3.5. Results Validation with the Stakeholders

Based on our discussions with farmers and data collection from literature, we assessed the performance of existing RWH with the evaluation tool. Then, the preliminary results were checked with our field observations and discussed with local farmers and experts. For instance, the RWH sites which scored 2 or lower (low suitability) had been abandoned and or most of their trees were dead. Whereas the sites that scored around 3 (medium suitability) showed well-maintained structures with healthy trees.

4. Discussion

Fifty-eight RWH sites (44 tabias and 14 jessr) in three sub-catchments were assessed and evaluated on their technical and economic performance as well on social aspects. Using our methodology, 65% of the assessed sites scored around 3 (medium suitability), 31% of the RWH sites got scores of about 2 (low suitability), and only 4%, two sites, scored 4 (high suitability). These results very accurately represent the real performance of each site—both overall and at individual criteria level based on the comparison of our observations and discussion with local farmers and experts. This suggests that the methodology developed is a valid way to assess the performance of RWH structures.

The percentage of each score for each criterion in all sites was shown in (Figure 4). Rainfall had the same score (Score 3) in all sites because of there was no big difference in rainfall pattern nor amount in three sub-catchments. This means the rainfall indicator has no significant impact on overall suitability between sites in our case study, but it can be very important in the comparison between sites in the larger areas [21] with a significant difference in rainfall. Moreover, significantly low score percentages were obtained by the design criteria, drainage length and cost, which was Score 1. For example, drainage length scored 1 for 48% of all sites. That means the distance between watercourses and RWH structures is big and the score would have been higher if these structures were built closer to the watercourse. If the RWH structures were located much closer to the watercourses, the contribution of drainage length to the overall RWH suitability would have been higher for our case study. Therefore, drainage flow has a significant impact on the performance of the RWH, which is not always the case for other types of RWH such as ponds, terraces, *etc.*

It is interesting to note that although the weight for climate criteria was higher than that for site characteristics criteria, 30% and 26% respectively (Figure 5),the latter received the highest scores in most of the sites in all three sub-catchments (Figures 6 and 7). This indicates that the sites are generally well selected for their purpose, and the site characteristics criteria had more impacts on the performance of RWH than other criteria such as climate, drainage and structure design. These results are similar to other studies, such as Al-Adamat [6] and Mbilinyi [16], who concluded that site characteristics are the most important criteria to be considered for design and implementation of RWH techniques.

Where RWH performance (suitability) was low, it was in most cases related to a shortcoming in the engineering design, lack of proper maintenance and the high cost of the water storage. The low performance of these RWH sites was confirmed by getting low scores of these criteria, as shown in Figure 4. The evaluation using our methodology clearly shows which criteria should be addressed to improve the performance of, for example, RWH structure design and storage capacity criteria. Due to the small storage area relative to the dyke size, the cost per cubic meter of water, especially in the jessour, was very expensive—such as jessour 10 and 15 in Sub-catchment 1. These results confirm that water harvesting structures with small storage capacity can ultimately be more expensive than large structures, as shown by Lasage, R., & Verburg, P.H. [4]. Therefore, if farmers can improve the dyke design and storage capacity area by following some basic engineering principles such as increasing storage area, constructing a regular spillway and providing periodic maintenance, they will be able to collect more water with less cost and keep the structure working for a longer period of time. Another example is the ratio between catchment size and cultivated area. Where this is not suitable, such as structures 11 and 21 in Sub-catchments 2 and 3, respectively, RWH structure performance can be improved by adapting the cultivated area to the effective area where the water is stored and adapting the crop type or cropping density (which determines the water requirement) to the amount of water stored.

In our methodology, two methods were applied (three levels and two levels of AHP hierarchy structure), and the results for both approaches were very similar. The consistency ratio for both methods was also similar and strong. Therefore, both methods are valid and provide reliable results. Both methods are simple to apply and easy to adapt the criteria in case of different RWH techniques and/or regions in order to cater to stakeholders' objectives. While either method can be used, it is recommended to apply Method 1 (three levels). In Method 1, the impact of any errors in scores (from expert opinion or calculations) will be reduced through the two-step calculation.

In most previous studies, the number of criteria are limited and are aimed primarily at the selection of suitable locations for RWH [1,14] and do not consider other factors or performance over time. In addition, many of those studies were mainly desktop studies using GIS and RS, without including stakeholders' objectives and constrains. Our study showed that socio-economic aspects play an important role in RWH suitability and performance. Thus, the inclusion of such criteria as occurs in our methodology is very important to the goal of meaningful information for improving current RWH effectiveness as well as planning for future structures.

A key precondition for the methodology was that it can be widely applied for different RWH techniques in different regions. In this regard, the structure of the methodology allows it to be easily adapted and applied to different RWH techniques and social-economic settings by simply changing the criteria selected. In addition, the case study showed that it is very possible to select criteria that are easy to assess and still provide accurate results without the need for complex analysis. This keeps the time investment and costs required within reasonable limits.

While Al-Adamat 2008 [6], Jabr 2005 [37] and Mbilinyi 2005 [33] showed that MCA provides a rational, objective and non-biased method for identifying suitable RWH sites, our study demonstrates that combining MCA and expert opinion in a consistent way allows assessment and evaluation of RWH techniques beyond simply site selection. Site conditions and RWH structure performance are likely to change over time, especially in light of predicted climate change. Therefore, a methodology such as ours, which allows evaluation of the performance of current and potential RWH projects, and identification of necessary improvements, is of great value.

An important consideration in the application of our methodology that warrants mention is the establishment of the scores/weighting for each criterion. As this depends on expert opinion [24,27], it is essential to use several experts and take into consideration their area of specialty when analyzing and using their inputs.

5. Conclusions

An evaluation and decision support methodology/tool was developed and tested for assessment of the overall performance of existing RWH and criteria affecting that performance. A single-objective AHP supported by GIS was put to the test in the Oum Zessar watershed of south-eastern Tunisia to assess the performance of 58 RWH structures (jessour/tabias) in three main sub-catchments. Engineering (Technical), biophysical and socio-economic criteria were determined, weighted and assessed in this study with input from experts and stakeholders. The main conclusions are:

(a) The methodology provides an accurate evaluation of RWH performance when compared with the field investigations;
(b) The methodology provides a good insight into where in the system improvements are needed for a better performance;

(c) In the case study, most sites showed low suitability scores for the criteria structure design, drainage flow and cost, which resulted in a low score on the overall performance of RWH;

(d) Site characteristics criteria (both overall and individual criterion) play a more important role in the overall suitability than other criteria;

In addition, the methodology can be used to pre-evaluate potential new RWH projects, increasing the chances for good long-term performance. This case study application of our methodology confirmed that it is a highly flexible and applicable tool for the evaluation and improvement of RWH structures, and can employ many different, important and easy to access criteria and indicators in the assessment of different RWH techniques. The time and cost required in using this methodology are also low, making it accessible to the local RWH managers/communities.

To further validate the applicability of the methodology, it needs to be tested in different regions and with different RWH techniques. Moreover, the criteria related to socio-economic suitability/ performance (*i.e.*, ownership, family size, *etc.*) deserve further investigation. These suggestions will increase the reliability and applicability of our methodology so that it can be used for assessing the performance of existing and new planned RWH structures in any region. This new, scientifically-based evaluation and decision support tool provides a basis on which designers and decision makers can build efficient RWH systems to meet the objectives and needs of the communities in water-scarce regions.

Acknowledgments: This study has been conducted in the framework of the cooperation between The Higher Committee For Education Development in Iraq (HCED) and Wageningen University (The Netherlands) and was done under the European Union's Seventh Framework Programme (FP7/2007-2013,WAHARA project). Field works have been carried out in collaboration with the *Institut des Régions Arides* (IRA) in Tunisia. Special thanks are due for Ammar Zerrim, Abdeladhim Mohamed Arbi and Messaoud Guied for the technical and field measurements assistance. Great thanks to Demie Moore in the Soil Physics and Land Management Group, Wageningen University, for the English reviewing and supporting.

Author Contributions: The manuscript was primarily written by Ammar Adham. Each one of the authors contributed to this work, Ammar Adham collected required data, assisted by Mohamed Ouessar in the field works/measurements and the data analysis. Ammar Adham and Michel Riksen developed the structure of the study, the manuscript, results, discussion and concluding remarks. Coen Ritsema supervised and guided this work. All authors contributed to the development of the approach, editing multiple drafts, and offering comments and corrections.

Conflicts of Interest: The authors declare no conflict of interest.

References

1. Kahinda, J.M.; Lillie, E.S.B.; Taigbenu, A.E.; Taute, M.; Boroto, R.J. Developing suitability maps for rainwater harvesting in South Africa. *Phys. Chem. Earth Parts A/B/C* **2008**, *33*, 788–799. [CrossRef]

2. Boers, T.M.; Ben-Asher, J. A review of rainwater harvesting. *Agric. Water Manag.* **1982**, *5*, 145–158. [CrossRef]

3. Kahinda, J.M.; Rockström, J.; Taigbenu, A.E.; Dimes, J. Rainwater harvesting to enhance water productivity of rainfed agriculture in the semi-arid Zimbabwe. *Phys. Chem. Earth Parts A/B/C* **2007**, *32*, 1068–1073. [CrossRef]

4. Lasage, R.; Verburg, P.H. Evaluation of small scale water harvesting techniques for semi-arid environments. *J. Arid Environ.* **2015**, *118*, 48–57. [CrossRef]

5. Biazin, B.; Sterk, G.; Temesgen, M.; Abdulkedir, A.; Stroosnijder, L. Rainwater harvesting and management in rainfed agricultural systems in sub-Saharan Africa—A review. *Phys. Chem. Earth Parts A/B/C* **2012**, *47*, 139–151. [CrossRef]

6. Al-Adamat, R. GIS as a decision support system for siting water harvesting ponds in the Basalt Aquifer/NE Jordan. *J. Environ. Assess. Policy Manag.* **2008**, *10*, 189–206. [CrossRef]

7. Ammar, A.; Riksen, M.; Ouessar, M.; Ritsema, C. Identification of suitable sites for rainwater harvesting structures in arid and semi-arid regions: A review. *Int. Soil Water Conserv. Res.* **2016**. [CrossRef]

8. Ouessar, M.; Sghaier, M.; Mahdhi, N.; Abdelli, F.; de Graaff, J.; Chaieb, H.; Yahyaoui, H.; Gabriels, D. An integrated approach for impact assessment of water harvesting techniques in dry areas: The case of oued Oum Zessar watershed (Tunisia). *Environ. Monit. Assess.* **2004**, *99*, 127–140. [CrossRef] [PubMed]

9. Oweis, T.Y. Rainwater harvesting for alleviating water scarcity in the Drier environments of West Asia and North Africa. In Proceedings of the International Workshop on Water Harvesting and Sustainable Agriculture, Moscow, Russia, 7 September 2004; p. 182.

10. Mechlia, N.B.; Oweis, T.; Masmoudi, M.; Khatteli, H.; Ouessar, M.; Sghaier, N.; Anane, M.; Sghaier, M. *Assessment of Supplemental Irrigation and Water Harvesting Potential: Methodologies and Case Studies from Tunisia;* ICARDA: Aleppo, Syria, 2009.

11. De Winnaar, G.; Jewitt, G.P.W.; Horan, M. A GIS-based approach for identifying potential runoff harvesting sites in the Thukela River basin, South Africa. *Phys. Chem. Earth Parts A/B/C* **2007**, *32*, 1058–1067. [CrossRef]

12. Kadam, A.K.; Kale, S.S.; Pande, N.N.; Pawar, N.J.; Sankhua, R.N. Identifying Potential Rainwater Harvesting Sites of a Semi-arid, Basaltic Region of Western India, Using SCS-CN Method. *Water Resour. Manag.* **2012**, *26*, 2537–2554. [CrossRef]

13. Ramakrishnan, D.; Bandyopadhyay, A.; Kusuma, K.N. SCS-CN and GIS-based approach for identifying potential water harvesting sites in the Kali Watershed, Mahi River Basin, India. *J. Earth Syst. Sci.* **2009**, *118*, 355–368. [CrossRef]

14. Ziadat, F.; Oweis, T.; Mazahreh, S.; Bruggeman, A.; Haddad, N.; Karablieh, E.; Benli, B.; Zanat, M.A.; Al-Bakri, J.; Ali, A. *Selection and Characterization of Badia Watershed Research Sites;* International Center for Agricultural Research in the Dry Areas (ICARDA): Aleppo, Syria, 2006.

15. Elewa, H.H.; Qaddah, A.A.; El-feel, A.A.; Brows, J.; St, T.; Nozha, E.; Gedida, E.; Alf-maskan, P.O.B. Determining Potential Sites for Runoff Water Harvesting using Remote Sensing and Geographic Information Systems-Based Modeling in Sinai. *Am. J. Environ. Sci.* **2012**, *8*, 42–55.

16. Mbilinyi, B.P.; Tumbo, S.D.; Mahoo, H.F.; Mkiramwinyi, F.O. GIS-based decision support system for identifying potential sites for rainwater harvesting. *Phys. Chem. Earth Parts A/B/C* **2007**, *32*, 1074–1081. [CrossRef]

17. Ouessar, M.; Bruggeman, A.; Mohtar, R.; Ouerchefani, D.; Abdelli, F.; Boufelgha, M. Future of drylands—An overview of evaluation and Impact Assessment Tools for water harvesting. In *The Future of Drylands;* Lee, C., Schaaf, T., Eds.; Springer: Berlin, Germany, 2009; pp. 255–267.

18. Jothiprakash, V.; Sathe, M.V. Evaluation of rainwater harvesting methods and structures using analytical hierarchy process for a large scale industrial area. *J. Water Resour. Prot.* **2009**, *1*. [CrossRef]

19. Mahmoud, S.H. Delineation of potential sites for groundwater recharge using a GIS-based decision support system. *Environ. Earth Sci.* **2014**, *72*, 3429–3442. [CrossRef]

20. Critchley, W.; Siegert, K. *Water Harvesting;* FAO: Rome, Italy, 1991.

21. Mahmoud, S.H.; Alazba, A.A. The potential of *in situ* rainwater harvesting in arid regions: Developing a methodology to identify suitable areas using GIS-based decision support system. *Arab. J. Geosci.* **2014**, *72*, 3429–3442. [CrossRef]

22. Ouessar, M. *Hydrological Impacts of Rainwater Harvesting in Wadi Oum Zessar Watershed (Southern Tunisia);* Ghent University: Ghent, Belgium, 2007.

23. Ouessar, M.; Zerrim, A.; Boufelgha, M.; Chniter, M. Water harvesting in south-eastern Tunisia: State of knowledge and challenges. In *Water Harvesting in Mediterranean Zones: An Impact Assessment and Economic Evaluation, Proceedings of EU Wahia Project Final Seminar in Lanzarote, 2002;* Wageningen University: Wageningen, The Netherland, 2002, Volume 40, pp. 13–24.

24. Adamcsek, E. The Analytic Hierarchy Process and Its Generalizations. Ph.D. Thesis, Eotvos Lorand University, Budapest, Hungary, 2008.

25. Ying, X.; Zeng, G.M.; Chen, G.Q.; Tang, L.; Wang, K.L.; Huang, D.Y. Combining AHP with GIS in synthetic evaluation of eco-environment quality-A case study of Hunan Province, China. *Ecol. Modell.* **2007**, *209*, 97–109. [CrossRef]

26. Saaty, T.L. Decision making with the analytic hierarchy process. *Int. J. Serv. Sci.* **2008**, *1*, 83–98. [CrossRef]

27. Al-Adamat, R.; Diabat, A.; Shatnawi, G. Combining GIS with multicriteria decision making for siting water harvesting ponds in Northern Jordan. *J. Arid Environ.* **2010**, *74*, 1471–1477. [CrossRef]

28. Tumbo, S.D.; Mbilinyi, B.P.; Mahoo, H.F.; Mkiramwinyi, F.O. Determination of suitability levels for important factors for identification of potential sites for rainwater harvesting. In Proceedings of the 7th WaterNet-WARFSA-GWP-SA Symposium, Lilongwe, Malawi, 1–3 November 2006; p. 15.

29. De Winnaar, G.; Jewitt, G.P.W.; Horan, M. A GIS-based approach for identifying potential runoff harvesting sites in the Thukela River basin, South Africa. *Phys. Chem. Earth Parts A/B/C* **2007**, *32*, 1058–1067. [CrossRef]

30. Mekdaschi Studer, R.; Liniger, H. *Water Harvesting: Guidelines to Good Practice*; Centre for Development and Environment (CDE): Bern, Switzerland, 2013.

31. Isioye, O.A.; Shebe, M.W.; Momoh, U.O.; Bako, C.N. A Multi Criteria Decision Support System (MDSS) for Identifying Rainwater Harvesting Site (S) in Zaria, Kaduna State, Nigeria. *Int. J. Adv. Sci. Eng. Tech. Res.* **2012**, *1*, 53–71.

32. Schiettecatte, W.; Ouessar, M.; Gabriels, D.; Tanghe, S.; Heirman, S.; Abdelli, F. Impact of water harvesting techniques on soil and water conservation: A case study on a micro catchment in southeastern Tunisia. *J. Arid Environ.* **2005**, *61*, 297–313. [CrossRef]

33. Mbilinyi, B.P.; Tumbo, S.D.; Mahoo, H.F.; Senkondo, E.M.; Hatibu, N. Indigenous knowledge as decision support tool in rainwater harvesting. *Phys. Chem. Earth Parts A/B/C* **2005**, *30*, 792–798. [CrossRef]

34. Oweis, T.Y.; Prinz, D.; Hachum, A.Y. *Rainwater Harvesting for Agriculture in the Dry Areas*; CRC Press: London, UK, 2012; p. 262.

35. Bamne, Y.; Patil, K.A.; Vikhe, S.D. Selection of Appropriate Sites for Structures of Water Harvesting in a Watershed using Remote Sensing and Geographical Information System. *Int. J. Emerg. Tech. Adv. Eng.* **2005**, *2025*, 270–275.

36. Nasr, M. *Assessing Desertification and Water Harvesting in the Middle East and North Africa: Policy Implications*; Zentrum f{ü}r Entwicklungsforschung-ZEF: No. 10; Center for Development Research: Bonn, Germany, 1999; p. 59.

37. Jabr, W.M.; El-Awar, F.A. GIS & analytic hierarchy process for siting water harvesting reservoirs, Beirut, Lebanon. *J. Environ. Eng.* **2005**, *122*, 515–523.

water

MDPI

Article

Rooftop Rainwater Harvesting for Mombasa: Scenario Development with Image Classification and Water Resources Simulation

Robert O. Ojwang [1], Jörg Dietrich [2,*], Prajna Kasargodu Anebagilu [2], Matthias Beyer [3] and Franz Rottensteiner [4,*]

[1]   Coast Water Services Board, Mikindani Street, Off Nkurumah Road, 90417-80100 Mombasa, Kenya; robertojwang@yahoo.com

[2]   Institute of Hydrology and Water Resources Management, Leibniz Universität Hannover, Appelstr. 9A, 30167 Hannover, Germany; prajna@iww.uni-hannover.de

[3]   Department B2.3, Groundwater Resources—Quality and Dynamics, Federal Institute for Geosciences and Natural Resources (BGR), Stilleweg 2, 30655 Hannover, Germany; matthias.beyer@bgr.de

[4]   Institute of Photogrammetry and GeoInformation, Leibniz Universität Hannover, Nienburger Str.1, 30167 Hannover, Germany

*   Correspondence: dietrich@iww.uni-hannover.de (J.D.); rottensteiner@ipi.uni-hannover.de (F.R.); Tel.: +49-511-762-2309 (J.D.); + 49-511-762-2483 (F.R.)

Academic Editor: Ataur Rahman

Received: 28 February 2017; Accepted: 15 May 2017; Published: 20 May 2017

Abstract: Mombasa faces severe water scarcity problems. The existing supply is unable to satisfy the demand. This article demonstrates the combination of satellite image analysis and modelling as tools for the development of an urban rainwater harvesting policy. For developing a sustainable remedy policy, rooftop rainwater harvesting (RRWH) strategies were implemented into the water supply and demand model WEAP (Water Evaluation and Planning System). Roof areas were detected using supervised image classification. Future population growth, improved living standards, and climate change predictions until 2035 were combined with four management strategies. Image classification techniques were able to detect roof areas with acceptable accuracy. The simulated annual yield of RRWH ranged from 2.3 to 23 million cubic meters (MCM) depending on the extent of the roof area. Apart from potential RRWH, additional sources of water are required for full demand coverage.

Keywords: Mombasa; roof rainwater harvesting; water supply; water demand; integrated water resources management; WEAP

---

1. Introduction

Rainwater harvesting is a technique used to collect and store rainwater e.g., from buildings, rock catchments, and land or road surfaces. The authors of [1–3] describe rainwater harvesting to be a dominant contributor for sufficing urban water demand. Rooftop rainwater harvesting (RRWH) refers to the collection and storage of water from rooftops [4]. The level of expertise required is low and ownership can be at a household level, making it easily acceptable to many people [1,5,6]. RRWH can support the water supply in almost any place either as a sole source or by reducing stress on other sources through water savings. The authors of [7] observed that the most important feature of RRWH at a domestic level is its ability to deliver water to households "without walking". This is particularly important in developing countries where women and children have to walk over long distances to fetch water. RRWH can be one aspect of the adaptation of water supply systems to climate change [8]. In Sub-Saharan Africa, the reliability of RRWH systems for domestic water supply can be improved by the consideration of rainfall characteristics, e.g., the number of events above a certain threshold, wet spells, etc., and improved technical design [9].

The quantity and quality of the harvested rainwater greatly depend on the type of roofing material used. Hard surfaces like iron, concrete, and tiles produce the highest amount of collected water because they have high runoff coefficients. A study by [10] showed that galvanized steel yielded the best quality rainwater that met the WHO (World Health Organization) drinking water guidelines for chemical, physical, and biological parameters. Furthermore, the slope of the roof has considerable influence on the roof runoff [11]. The "first flush" is a criteria often used for the design of RRWH systems. It describes the amount of initial rainfall, which is needed after a dry period to remove contaminants such as particles, dirt, bird droppings, and insect bodies from the roof and the gutter. First flush diverters need to be incorporated into the system in order to protect the water quality in the collection tank from contamination [1,12]. Due to first flush diversion, the quantitative yield of RRWH systems is reduced.

The amount of water harvested can be calculated using the rational method commonly used for very small urban catchments if the roof area, amount of rainfall, and the roof runoff coefficient are known. The runoff coefficient represents losses due to evaporation and leakages [1,13]. A study in Jordan showed that the potential of rainwater harvesting is about 15.5 MCM per year, which allowed potable water savings ranging from 0.3% to 19.7% in 12 administrative units [1]. In the UK, an average water saving efficiency (ET) of 87% over a period of 8 months was reported for the total toilet flushing demand of an office building using RRWH. ET is the percentage ratio of rainwater supplied to the total estimated demand [14]. For Iran, [15] reported supply rates of 75% but for different durations of 40–75% of the time, depending on climatic conditions.

The use of collected rainwater for domestic purposes (tertiary uses like gardening) is a major component of water supply in the rural areas of South Africa, with 96% of 34,000 RWH tanks being located in rural regions [8]. When harvested water is used as potable water, the quality becomes paramount. The authors of [6] showed that most people in Zambia expressed interest in RRWH, but were concerned about its quality. Several other studies have shown that rainwater usually meets the WHO standards for physical and chemical parameters but may fail regarding the biological parameters, i.e., fecal and total coliform counts [10,16–18].

The key parameters for estimating RRWH water yield are roof area and rainfall. Another major factor is the estimation of roof areas, which is difficult especially for unplanned city areas. The authors of [10] list several methods:

- Sampling: representative samples of rooftops are obtained and extrapolated to the total area. This method is suitable for estimating roof areas for large areas;
- Multivariate sampling: correlations are drawn between additional variables (e.g., population) and roof area;
- Complete census: gives the most accurate results but involves the computation of the entire area of the rooftops in the area of interest by using statistical information like floor area, number of floors, and number of housing units;
- Digitization or image classification tools can be used from remotely sensed high-resolution images to compute the roof areas with a Geographical Information System (GIS).

The authors of [1] used a complete census to estimate roof areas in Jordan. Available information on the dwelling units such as different types of units, number of units per type, and average area per type were used to estimate the roof area in each of the governorates. The same approach was applied in another study in Seoul, South Korea to determine the city's rooftop photovoltaic (PV) potential [19]. The results were validated by using automated vector detection software. In a study of the informal settlement of Diepsloot near Johannesburg, South Africa, an automatic feature extraction (image classification) method was applied using aerial satellite imagery to extract roof areas with

80% accuracy [20]. Similarly, in the Kibera slums, Nairobi, feature extraction was successfully used to estimate rooftop areas, which was then used to derive the population in the slum [21].

The rooftop area is a crucial input for the incorporation of RRWH systems into water resource system models, which can be applied for the quantitative planning of water resources and the development of water policies. WEAP (Water Evaluation and Planning System), developed by the Stockholm Environment Institute (SEI, Stockholm, Sweden), is a planning tool extensively used in integrated water resources management. It is both a model for simulating water systems in an integrated manner (natural and manmade components/infrastructures) and a policy oriented decision-support system (DSS) [22]. Demand and supply sites are considered concurrently. WEAP follows the principle of the "scenario-based gaming approach" that has been developed to reduce water demand-supply conflicts within the area of interest [23]. The model uses scenarios to promote stakeholder involvement in the entire water resources planning and decision making process. The water resources system in WEAP is represented by demand sites, supply sites, catchments, withdrawal points, transmission links, wastewater treatment, environmental needs, and the generation of pollution. Depending on the need and data availability, WEAP simulates several aspects such as sectoral water demands, water allocation rights and priorities, ground and surface water flows, reservoir operations, and the assessment of vulnerability and cost-benefit analysis, amongst others.

The authors of [24] tested WEAP's demand management scenario evaluation in the water-stressed Olifants river basin, South Africa, and [25] applied WEAP in the study of the Upper Ewaso Ng'iro North Basin, Kenya to balance the water requirements of competing users against the available water resources in the basin. The study found that the use of WEAP improved the complex system of demand-supply of the basin. Applications of WEAP in the development of RRWH strategies have not been documented so far.

In this study, WEAP is applied to develop a variety of future scenarios of RRWH for the city of Mombasa by using projections of population growth and climate change. This research contributes to closing the gaps in the current methods of investigating RRWH by combining remote sensing with water resources modelling. The main objectives of the study are as follows:

- Determination and discrimination of rooftop areas and different roof types from high resolution satellite images;
- Setup and parameterization of an extended WEAP model with an implemented simple RRWH scheme for large scale planning;
- Implementation of future scenarios in WEAP and evaluation of their implications and potential for long-term management of the urban water supply.

2. Materials and Methods

2.1. Study Area and Data

Mombasa City is the second largest city in Kenya with an estimated 1.1 million inhabitants and a land area of 229.7 km^2. The city is located in the southern part of Kenya and is divided into four main areas: Island, South Mainland, West Mainland, and North Mainland (Figure 1).

The city experiences a tropical climate, which is hot and humid throughout the year with a mean daily minimum and maximum temperature of 22 °C and 30 °C, respectively. Annual precipitation is 1024 mm (data from 1984 to 2013 by the Kenya Meteorological Department). There are two rainy seasons. For the long rainy season between April and July, the average monthly rainfall is 134 mm. Between October and December, there are short rains with 100 mm average monthly rainfall. During the dry season, the average monthly rainfall is 37 mm with several months without rain. These variations in climatic conditions are attributable to the S-E and N-E monsoon winds and oceanic factors. The average monthly areal precipitation over the catchment was computed as the arithmetic mean of 10 rain gauge stations within the catchment. Annual potential evaporation exceeds the rainfall by magnitudes, which results in freshwater deficits during the dry periods. Reference evapotranspiration (ET0) data were obtained from the FAO (Food and Agriculture Organization of the United Nations) CLIMWAT 2.0 database using Moi International Airport, Mombasa station located at 4.03° S and 39.61° E. According to the Kenya Integrated Water Resources Management and Water Efficiency Plan, the country's annual average water availability in 2002 was 647 m^3 per capita. This ranks Kenya as a water scarce country according to [26], where below 1700 m^3/capita/year means water stress, below 1000 m^3/capita/year means water scarcity, and below 500 m^3/capita/year means absolute water scarcity. The situation in Kenya is predicted to worsen to 235 m^3 per capita by the year 2020 [27].

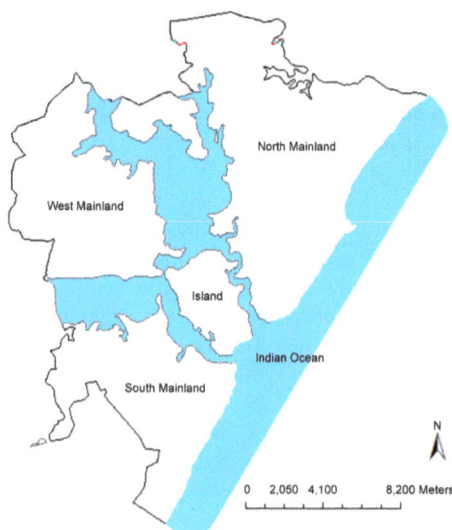

Figure 1. Map of Mombasa City showing the four main zones in the study area.

For future climate projections, downscaled precipitation and temperature data from the CIMP5 Global Climate Models (GCM) were used for the RCP 4.5 and RCP 8.5 pathways. The Intergovernmental Panel on Climate Change (IPCC) in its Climate Change 2014 Synthesis Report uses Representative Concentration Pathways (RCPs) to describe four different pathways of greenhouse gas concentration in the atmosphere, named after the change in radiative forcing in W/m^2 compared to pre-industrial times. In the case of RCP 4.5, the global mean surface temperature is likely to increase from 1.1 K to 2.6 K for 2081–2100 relative to 1986–2005.

The CIP datasets provide time series data (observed and downscaled projections) for different climatic variables for weather stations across Africa. The projections are compiled of downscaled products of 11 GCMs, summarized in Table 1. From the long-term historical (1971–2000) monthly ET0 data, future values can be estimated based on temperature changes. The authors of [28] suggested that for each degree rise in temperature, there is a corresponding 5% increase in ET0. The historical ET0 values were then corrected for both RCP 4.5 and RCP 8.5 to obtain future estimations.

In order to address model uncertainty, multi-model ensemble averages can be used; however, spatiotemporal information, especially of extreme events, can be lost [29,30]. The evaluation of bias (Table 1) and root mean square error (RMSE, not shown) between the downscaled RCP 4.5 and RCP 8.5 data (monthly series) and the observed historical data for the period from 1984 to 2013 showed that the bias for the GFDL-ESM2G model is lower than the ensemble bias, but the RMSE is lowest for the ensemble. Furthermore, it is not clear if the bias is non-stationary. Thus, using the model ensemble is preferable over using one single model. Multi-model means were used to generate future monthly rainfall series in this study.

Table 1. Details of the GCM models used and their bias (%) in the height of precipitation for the historical period from 1984–2013.

Model Name	RCP 4.5	RCP 8.5	Climate Modelling Institution/Centre
MIROC-ESM	−12.70	−15.70	National Institute for Environmental Studies, Japan
CNRM-CM5	−8.00	−8.10	Centre National de Recherches Meteorologiques, France
CAN-ESM2	−3.80	−2.30	Canadian Centre for Climate Modelling and Analysis
FGOALS-S2	−13.80	−19.70	Institute of Atmospheric Physics, Chinese Academy of Sciences
BNU-ESM	−16.00	−15.00	Beijing Normal University
MIROC5	9.20	8.90	National Institute for Environmental Studies, Japan
GFDL-ESM2G	0.40	2.70	Geophysical Fluid Dynamics Laboratory, USA
MIROC-ESM-CHEM	−15.50	−15.40	National Institute for Environmental Studies, Japan
GFDL-ESM2M	−1.20	−1.70	Geophysical Fluid Dynamics Laboratory, USA
MRI-CGCM3	26.90	22.20	Japan Meteorological Agency
BCC-CSM1-1	−10.90	−10.70	Beijing Climate Centre
Ensemble Average	−4.10	−5.00	-

Only domestic water demand was considered in this study. In order to estimate the per capita water consumption (water use rate), the recommended amounts for different categories according to the practice manual for water supply services [31] together with poverty level data for Mombasa City [32] were used (Table 2). The estimated water use rate is around 116 L per capita per day (LCPD).

Table 2. Domestic water use rates for Mombasa City.

Category	Persons	Water Use Rate	Remarks
	Percent	LCPD	
High Class Houses (HCH)	5.00%	250	
Medium Class Houses (MCH)	57.40%	150	Poverty level was 37.6% in 2013,
Low Class Houses with individual connections (LCH_IC)	18.80%	75	remaining 62.4% assumed as 5% for HCH and 57.4% for MCH
Low Class Houses without individual connections (LCH_WIC)	18.80%	20	
Weighted		116	-

Assuming the population will continue to grow at the average rate of 3.2% experienced between 1999 and 2009 [33], the population is expected to rise to 2.2 million people by 2035, leading to an increased water demand from 150,000 m³/day to 320,000 m³/day (Figure 2). The existing water supply to the city is 102,000 m³/day from Mzima springs, Tiwi boreholes, Marere Springs, and the Baricho Wellfield managed by the Coast Water Services Board (CWSB). Considering system losses of around 47%, the current demand coverage is low [34,35].

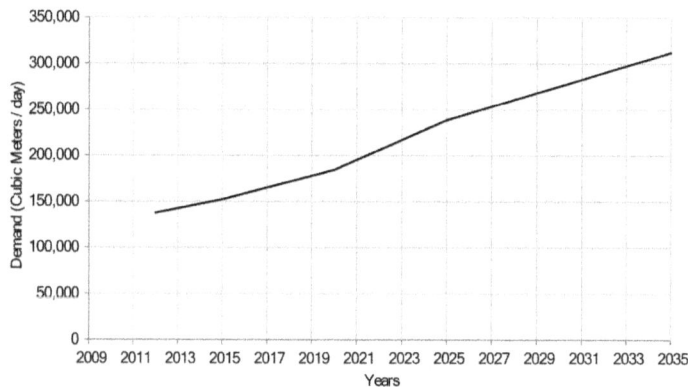

Figure 2. Water demand projections for Mombasa City [34].

The water supply master plan [34] identified several projects to help bridge the gap, mostly by the expansion of the Baricho Wellfield and Mzima Springs to full capacity, the construction of the 228,000 m^3/day Mwache Dam, and the acquisition of the Mkurumudzi dam. The plan considered RRWH, desalination, and wastewater reuse as alternative sources of water even though no estimates of quantities are available at this point. The biggest challenge is the large amount of investment capital required. It is further doubtful if the projects will fully cover the rising demand upon completion or if continued unsustainable extraction may lead to depletion. As a stopgap measure, the Mombasa Water and Sanitation Company (MOWASCO, Kenya) adopted water rationing of 6 h of supply per day [35]. Consequently, around 13,000 households use individual boreholes and hand-dug wells to supplement the conventional piped water supply [32]. Seawater intrusion due to over extraction may soon be a problem.

2.2. Overview of the Methodology

This study involved the following key steps: (i) data collection; (ii) image classification to estimate roof areas within Mombasa City; (iii) using estimated areas to calculate the rainwater harvesting potential and (iv) future scenario analysis with WEAP. In the scenario analysis, impacts of climate change on rainwater harvesting were also incorporated by using an ensemble of projections as shown in Table 1. The overall procedure and workflow is shown in Figure 3.

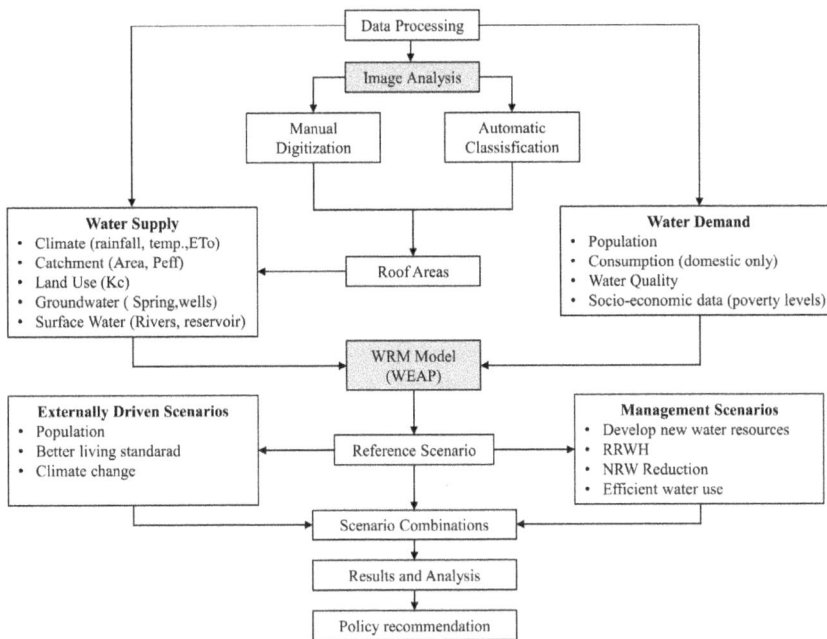

Figure 3. Methodological framework and workflow.

2.3. Roof Area Estimation

Roof areas were estimated manually by digitization and via automatic image classification using the ArcGIS® software by ESRI (Environmental Systems Research Institute, Redlands, CA, USA). The commonly used roofing materials in the city (tile, iron, and concrete making up 90% of the roof area) were considered. The following criteria were used to select the satellite images for this study from the Digital Globe foundation data for 2010 to 2014: (i) spatial resolution (high); (ii) cloud cover (low); and (iii) spatial coverage (extent of the study area covered). The best images in terms of spatial resolution were available from the satellite WorldView2 (WV2). Two WV2 images, which covered the whole study area, were selected for further processing and analysis (Table 3).

For manual digitization, only planned areas within the study area were targeted (referred to subsequently as the "control area") due to the ease of roof type identification for digitization. The other reason was that unplanned congested areas might not be suitable for RRWH since they lack the required space. Furthermore, the inhabitants may not be able to afford the system. Figure 4 shows the difference between planned and unplanned areas within the city.

Table 3. Selected images for processing (digitization and classification). The spatial resolution is after pan-sharpening.

Sensor Name	Acquisition Date	Cloud Cover (%)	Multispectral Bands	Off-Nadir (°)	Spatial Resolution (m)
WV-2	29/11/2013	0.6	8	23.10	0.5
WV-2	15/08/2013	8.9	8	6.60	0.5

(a)   (b)

Figure 4. Planned areas (**a**) and unplanned areas (**b**) (informal settlements).

Automatic image classification was used to estimate roof areas for the whole city. The performance of the automatic classification was assessed using the results from the manual digitization for the common areas covered by both techniques. In this study, supervised classification using the Gaussian maximum likelihood (ML) method was adopted [36]. Six classes were used: tiled roofs, iron roofs, concrete roofs, vegetation, roads, and ground. In ML classification, the user has to provide training areas for each class. In the training phase, the parameters of a Gaussian mean and covariance matrix are estimated from the image feature vectors of each class. The feature vector of an image pixel collects all the spectral values observed at that pixel. In the classification phase, for each pixel to be classified, the feature vector of that pixel is used to compute the Gaussian probability density for each class using the parameters estimated in the training phase. This value is referred to as the likelihood for the feature vector to belong to the respective class. The pixel is assigned to the class of maximum likelihood.

2.4. The WEAP Model for Mombasa City

2.4.1. Conceptual Model Scheme

The main water infrastructure of the city of Mombasa was implemented into WEAP as a conceptual model (Figure 5). Subsequently, a description of the demand and supply elements was incorporated into the model:

(a)   Demand site: Even though six different demand sites have been shown in the model (Mombasa City, Malindi Town, Kilifi Town, Kwale Town, Mariakani Town, and Voi Town), the study is focused only on Mombasa City and the rest are used to provide a complete picture of the sharing of water resources in the Coastal region.

(b)   Water sources:

(i)   Current situation:

○   The city receives water from Mzima Springs, Baricho boreholes, Marere Springs, Tiwi-Likoni boreholes, and individual dug-out wells.

○   The rivers of Marere, Mwache, Sabaki, and Rare are some of the rivers that flow around Mombasa City. However, currently there is no abstraction from these rivers.

(ii)   Future (presented in the model):

○   The head flow generated from the Mwache catchment feeds the Mwache River. The Mwache Reservoir is expected to supply water from 2020.

○ The rooftop areas are implemented as five catchment nodes, corresponding to the roof areas for each of the four zones in Mombasa, namely North Mainland (NML), South Mainland (SML), West Mainland (WML), Island, and new buildings to be constructed in the future. The water, which is collected from the rooftops of these five catchments, is directed into one reservoir "RRWH", which is modelled as a local reservoir.

○ Operation of Mkurumudzi Dam in supplying water to Mombasa is expected to start from 2030 [34].

○ Return flows are not considered in the WEAP model because the city mainly depends on onsite wastewater disposal methods such as pit latrines, cesspits, and septic tanks that do not allow any return flows to the rivers, and the sewer system of the city drains to the Indian Ocean. The two wastewater treatment plants, Kizingo and West Mainland, serve a very small population and also discharge directly to the Indian Ocean.

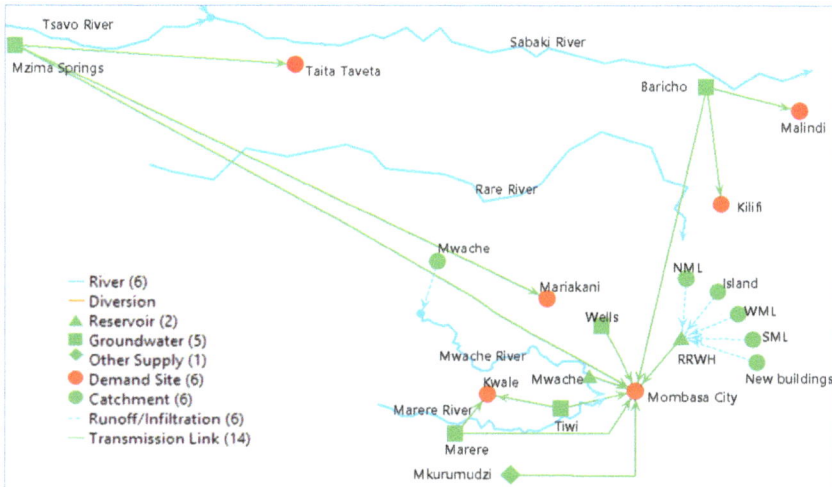

Figure 5. Conceptual model for Mombasa city (not drawn to scale).

2.4.2. Catchment and RRWH Implementation

The FAO rainfall runoff (simplified coefficient) method as implemented in WEAP was used in this study to simulate both natural and RRWH runoff generation from the catchments. This simple method calculates runoff by subtracting evapotranspiration from precipitation. The effective precipitation parameter ranges between 0% and 100% with 0% indicating that all precipitation produces runoff while 100% means that all precipitation is available for evapotranspiration.

The Mwache catchment covers approximately 2250 km^2 and the main vegetation in the catchment are deciduous forest and dry grasslands covering 38% and 62% of the area, respectively [37]. The FAO simplified method uses a crop coefficient K_c, which is relative to the reference crop for a particular land class type. The FAO Paper 56 for Irrigation and Drainage recommends K_c values for deciduous trees on grass between 0.8 and 0.9 and for deciduous trees with bare ground between 0.3 and 0.4. Assuming that dry grasslands are similar to bare ground, the K_c value was set to 0.5, while the effective precipitation was set to 65%. For the Mwache Reservoir, a storage capacity of 118.7 MCM was used, while the volume elevation curve and estimated monthly average reservoir evaporation were obtained from [37].

The five rooftop catchments represent the total rooftop area of each of the four zones of Mombasa with existing buildings (summarized as one catchment each) and one zone with new buildings. The size of the RRWH catchment area was determined from the roof identification by image classification depending on the scenario of roof usage considered. The precipitation of the RRWH catchments was corrected for the first flush loss occurring at the beginning of rainfall. According to the Texas Manual on Rainwater Harvesting [38], the first flush loss amounts to 1 Gallon (1 Gallon = 3.785 Liters) for every 100 Square feet (1 Square foot = 0.092 Square meters) of roof area, which is equivalent to 0.4 mm of rainfall. The average monthly number of rainy days from the observed data from 1984 to 2013 was used to estimate the average monthly first flush loss, which was then subtracted from the precipitation input. The runoff coefficients for the different roof materials were taken from [7], using 0.9 for iron and 0.8 for tile and concrete. Mombasa roofs have different slopes, hence there is no characteristic pitch. The roof pitches could not be obtained from the satellite images used in this study. Stereo images were not available, and an exhaustive ground observation was not possible. Similar to [1] and other studies, the roof angle could not be considered and average runoff coefficients were used based on the material alone.

The water collected from the rooftops of all five catchments is diverted into one reservoir in the WEAP model as a conceptual representation of the RRWH storage systems. This simplification was introduced because the potential of RRWH for additional water supply is investigated on a larger scale here. In reality, there can be centralized and/or decentralized solutions for the storage of the harvested water. The efficiency of different storage techniques, however, is not a subject of this study. The RRWH system storage was modelled using a single reservoir, which receives runoff from the five RRWH catchments (NML, SML, WML, Island, and new buildings) through runoff links. Since the target is to collect all of the rainwater, the storage capacity is set as unlimited. There is no evaporation from the RRWH reservoir, because it is simply a conceptual representation of closed storage tanks in WEAP. Thus, a fictitious volume elevation curve is used (1 m rise in level for each additional 1 m^3). Operation-wise, the top of the conservation zone is set to the storage capacity and no dead storage is provided since all of the water in the tank is assumed to be available for use depending on the demand requirements.

Flow from the supply sources to the demand sites was implemented using transmission links with the consideration of losses due to leakages. Losses were taken as 47% based on the latest Impact Report No. 7 [35] and 20% was assumed for the RRWH structures, representing the losses occurring mainly in the gutters, downpipes, and storage tanks.

2.4.3. Baseline Scenario

Due to the availability of suitable satellite images for roof area estimation and observed precipitation data, the year 2013 was chosen as the base year (Current Accounts Year in WEAP). For the last year of the scenarios, the year 2035 was selected in order to coincide with the last year of the Coast Water Services Board planning horizon and to be after the end of the Government of Kenya's Vision 2030 development blueprint. The temporal resolution of the model was set to yearly with monthly time steps. Parameters that might be subject to changes under different scenarios such as population growth rates, water use rates, non-revenue water (NRW) levels, or crop coefficients (K$_c$) were further implemented as "Key Assumptions". The data used in the current accounts is compiled in Table 4.

Table 4. Input data for current accounts.

Source	Parameters	Value	Reference
Roof	Crop coefficient, K_c	0.1	Lower than bare soil (0.3 from FAO Paper 56)
	Effective rainfall, Peff	Iron-10% Tile and concrete-20%	Reasonable assumptions
Groundwater sources: Baricho, Mzima, Tiwi, Marere, Ind. Wells	Storage Capacity Initial Storage (MCM) Max Withdrawal (MCM) Recharge	Unlimited 80, 82, 7.3, 7.3, 16 Same as initial storage 83, 405, 21, 15, 23	Reasonable assumptions Mumma and Lane, 2010; [34]; JBG Gauff Ingenieure, 1995; Samez Consultants, 2008; Sincat/Atkins Consultants, 1994; Fichtner/Wanjohi Consultants, 2014
Mwache Dam	Storage capacity (MCM) Evaporation Rate Effective rainfall, Peff Crop coefficient, K_c	118.7 Monthly rates 65% 0.5	[37]
Transmission	Loss in transmission links Loss in RRWH transmission	47% 20%	[35] Reasonable assumptions

2.4.4. Future Scenarios for Mombasa

The Reference Scenario inherits all of the information and data set up under the Current Accounts year (2013) and extends it over the entire timeframe (2014–2035) with no interventions to improve demand coverage. Here, the water supply remains at 102,000 m^3/day from the main sources and 29,143 m^3/day from the individual wells, with losses (NRW) of 47%. The per capita water consumption rate and population growth rate remain at 116 LCPD and 3.2% throughout the period, respectively.

Future scenarios for Mombasa are created to investigate the combined influence of (i) possible future changes of external factors, which are out of direct control of the water managers and which are uncertain, such as population growth, socio-economic dynamics, and climate change; and (ii) management decisions such as construction and expansion of more water sources, reduction of NRW, and implementation of RRWH. The growth assumptions for future scenarios are summarized in Table 5.

Table 5. Growth assumptions for Future Scenarios (External Factors).

Parameters	Value	Reference
High Population Growth (HPG) rate	4.2%	Kenya National Bureau of Statistics
Low Population Growth (LPG) rate	1.9%	(2009), BCEOM/Mangat (2011) and
Increased water consumption due to better standard of living	116 LPCD to 155 LPCD	Mombasa County (2014)

The scenarios driven by management decisions were: (i) Development of New Water Sources (NWS); (ii) Reduction of Non-Revenue Water Strategy (NRWS) where the NRW levels decrease from 47% to 20% to meet the Water Services Regulatory Board (WASREB) sector target; (iii) Efficient Water Use (EWU) where per capita consumption decreases from 116 LPCD to 93 LPCD; and (iv) RRWH scenarios where rainwater harvesting is practiced. The estimated allocated future flows to different parts of the city are shown in Table 6. The following five RRWH scenarios are investigated: (i) All existing roofs are used (RRWH_1); (ii) only new roofs are used (RRWH_2); (iii) selected existing buildings are used (RRWH_3); (iv) selected existing roofs and all new roofs are used (RRWH_4); and (v) all existing roofs and all new roofs to be used (RRWH_5). The different combinations of scenarios used in the model are shown in Table 7.

Table 6. Planned future flows to Mombasa City in m^3/day [34].

Source	Capacity	Current	Phase I		Phase II	Phase III	
		2014	2017	2020	2025	2030	2035
Baricho	175,000	60,000	82,000	55,805	106,594	80,395	80,395
Mzima	105,000	24,000	24,000	15,292	13,370	59,050	59,050
Marere	12,000	8000	8000	7135	6051	3173	3173
Tiwi	13,000	10,000	10,000	10,000	10,000	8662	8662
Mwache	228,000	0	0	95,595	102,859	145,838	145,838
Mkurumudzi	20,000	0	0	0	0	15,191	15,191

Table 7. Scenario combinations under the normal population growth rate.

Scenario Combination	Description
NWS/RRWH_4	Existing system with new water sources developed (NWS) and RRWH_4 (using all new roofs) implemented
NWS/NRWS	Existing system with NWS and non-revenue water (NRWS) reduction strategy implemented
RRWH_4/NRWS	Existing system with both RRWH_4 and NRWS strategy implemented
NWS/EWU	Existing system with new water sources developed and water use efficiency (EWU) improved
RRWH_4/NRWS/EWU	Existing system without new water sources developed but RRWH_4, NRWS, and EWU implemented
NWS/NRWS/EWU	Existing system with new water sources developed and NWRS and EWU implemented, but no RRWH_4
NWS/RRWH_4/NRWS/EWU	All strategies implemented (new water sources, RRWH, non-revenue water reduction, and efficient water use)

3. Results and Discussion

3.1. Determination of Rooftop Area

Figure 6 shows a visual comparison of a part of the classified image with the ground truth obtained by digitization (only roof classes shown). The visual assessment reveals that the classifier performed well in separating the different classes of roof material. The extent or degree of accuracy however cannot objectively be assessed by visual interpretation alone.

The results of manual digitization suggest that the selected control area has around 3 km^2 of suitable roofs for RRWH (based on non-congested planned areas). Table 8 summarizes the results of the manual digitization for the selected areas. Apart from representing one of the scenarios, the manual digitization was further used to assess the performance of the automatic classification technique. The total roof area within the city was found to be approximately 28 km^2 based on automatic image classification with 18.1 km^2 of iron, 5.6 km^2 tile, and 4.3 km^2 concrete (Table 9).

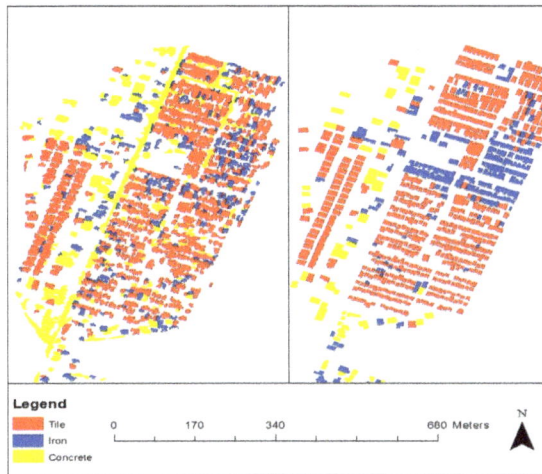

Figure 6. Comparison between manually digitized and automatic classification for a portion of the classified area (only roofing materials shown).

Table 8. Area of different roof materials (in m^2) resulting from the manual digitization for the selected control areas.

Zone	Tile	Iron	Concrete	Total
Island	610,339	114,854	298,186	1,023,379
North Mainland	1,139,521	313,385	418,285	1,871,192
South Mainland	42,367	34,158	2011	78,536
West Mainland	76,738	80,726	8439	165,902
Total	1,868,966	543,123	726,921	3,139,009

Table 9. Area of different roof materials (in m^2) resulting from automatic image classification for the whole investigation area.

Area	Tile	Iron	Concrete	Total
Island	1,686,856	2,874,256	1,300,597	5,861,709
North Mainland	2,115,315	6,550,999	1,934,685	10,600,999
South Mainland	66,500	3,644,231	116,438	3,827,169
West Mainland	1,759,221	5,044,778	944,178	7,748,177
Total	5,627,892	18,114,264	4,295,898	28,038,054

One common way to assess the accuracy of classification is to construct a confusion matrix [39]. The matrix indicates how the classifier confuses between the different classes. Three types of accuracies can be derived from the confusion matrix: overall accuracy (OA), user's accuracy (correctness, UA) and producer's accuracy (completeness, PA). OA is the percentage of pixels assigned to the correct class. For each class, UA gives the percentage of pixels assigned to that class that also belong to that class in the reference, whereas PA gives the percentage of pixels of that class in the reference that were also assigned to that class by the classification procedure.

Table 10 shows the confusion matrix generated from the image classification process. As suspected, the classifier has a significant problem in differentiating concrete from the background, for example, 3.6% of the background pixels are incorrectly labelled as concrete. Error analysis showed that the OA of the image classification was 88.6%. However, this high level of OA could be misleading because the background class constituted a larger percentage of the total area compared to the other three classes.

It was therefore necessary to check the producer's accuracy and user's accuracy for each class. The PA were 92.9%, 58.5%, 37.4%, and 54.3% for the background, tile, iron, and concrete, respectively. The UA on the other hand were 95.7%, 59.1%, 51.3%, and 22.2% for the background, tile, iron, and concrete, respectively. The high OA was due to the high values of PA and OA obtained for the background. When computing the confusion matrix for the differentiation of the background and roof (all materials), the PA is 93.0% for the background and 63.8% for the roof, and the UA is 95.6% for the background and 51.5% for the roof. Moreover, this reveals that the classifier had difficulties in differentiating some classes, especially tiles or concrete, from the background. This is largely because concrete roofs and roads have similar spectral properties. An idea to overcome this problem could be to use a Digital Surface Model, which can be derived from stereo images [40] by techniques such as semi-global matching [41]. Despite some further potential to improve the image classification, the OA of 88.6% indicates that the automatic classification was acceptable. Therefore, the results were used as input for the WEAP model.

Table 10. Confusion matrix based on selected pixels in the classified image.

		Class in Results (Automatic Classification) (All Values in %)					
		Background	Tile	Iron	Concrete	Sum	Completeness
	Background	83.3	1.7	1.0	3.6	89.6	92.9
Class in	Tile	1.8	2.8	0.1	0.2	4.8	58.5
Reference	Iron	1.0	0.2	1.2	0.8	3.2	37.4
(Digitized)	Concrete	1.0	0.1	0.0	1.3	2.4	54.3
	Sum	87.0	4.8	2.3	5.9	-	-
	Correctness	95.7	59.1	51.3	22.2	-	88.6

3.2. WEAP Scenarios

The projected water demand up to the year 2035 for Mombasa City under the different scenarios, namely the reference scenario, better living standards (BLS), high population growth (HPG), low population growth (LPG), and efficient water use (EWU), are shown in Figure 7.

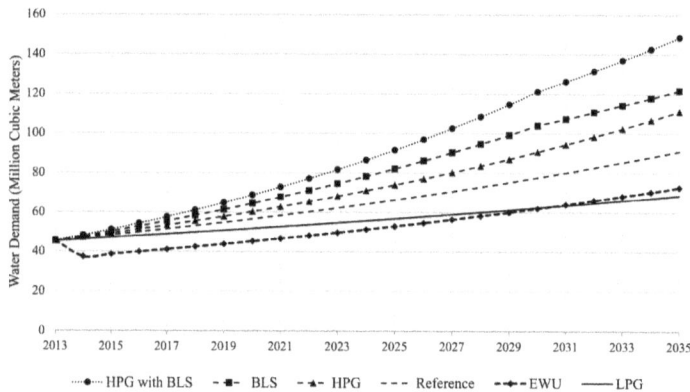

Figure 7. Annual Water demand projections in Million Cubic Metres (MCM) for different scenarios. Abbreviations: better living standards (BLS), high population growth (HPG), low population growth (LPG), efficient water use (EWU).

The simulated demand coverage in 2035 for Mombasa City with no interventions decreases from 54% in 2014 to 28% for the reference scenario. The demand coverage is lower under the HPG with BLS scenario (17%), BLS (21%), and HPG (23%) scenarios, but slightly improves in the EWU (35%) and LPG (37%) scenarios. As water demand coverage for Mombasa City continues to dwindle, it was necessary to investigate the impact of RRWH if practiced under different implementation scenarios. Figure 8 illustrates the amount of rainwater that can potentially be harvested under the various rainwater harvesting scenarios.

The results show that the potential of RRWH varies greatly between the different management scenarios. The inter-annual variability within all the scenarios is due to variations in precipitation over the years. The highest amount of rainwater can be provided under the scenario RRWH_5 (supply of over 28 MCM by 2035). The other scenarios RRWH_1 to RRWH_4 yield between 2.3 MCM and 20 MCM by 2035. These results suggest that the potential of RRWH in the city greatly depends on the strategy adopted by the city water management.

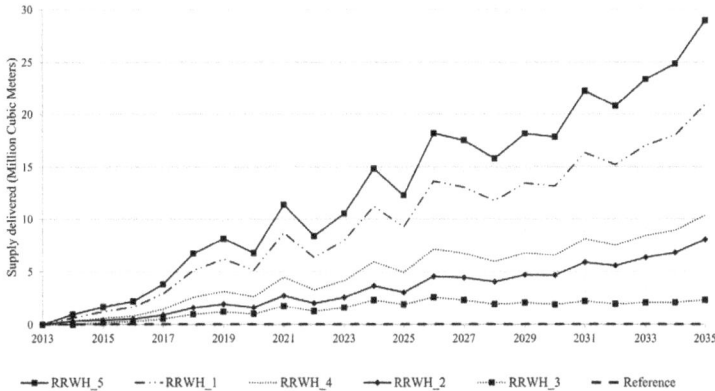

Figure 8. Additional water supply delivered under different RRWH scenarios for Mombasa City.

The average monthly demand coverage using the existing supply and RRWH as the only additional strategy is shown in Figure 9. For the reference scenario where no intervention was made to the existing system, the average monthly demand coverage for the whole 2014–2035 period is around 40%. The results clearly show that implementing RRWH improves the water supply situation. Based on the bimodal rainfall pattern in Mombasa City, higher demand coverage is achieved within the rainy months every year. In the month of May, which records the highest rainfall, the average monthly demand coverage increases to 46%, 53%, 58%, 75%, and 84% for RRWH_3, RRWH_2, RRWH_4, RRWH_1, and RRWH_5, respectively.

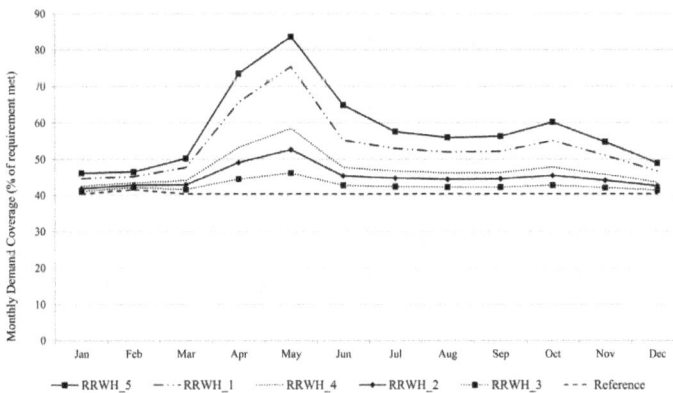

Figure 9. Average monthly demand coverage of RRWH combined with the existing system (2014–2035).

Apart from the RRWH strategies, the other possible management scenarios considered were the development and expansion of new water sources (NWS), efficient water use (EWU), and reduction of non-revenue water (NRWS). In terms of the water supply delivered for the whole period from 2013 to 2035, the responses of the system are presented in Table 11. The NWS scenario, which involves the expansion of existing water sources and the development of new ones, provides the best strategy, which increases the supply by 470 MCM throughout the whole study timeframe. The other strategies result in increases of the supply ranging between 34 MCM to 295 MCM over the same period. The UWE scenario does not increase the supply but reduces the demand, hence its effect can only be seen under demand coverage or unmet demand. Figure 10 indicates that no single strategy will completely solve the water scarcity in the city. Thus, a combination of different strategies is recommended.

Table 11. Total supply delivered under different management scenarios (2013–2035).

Supply/Demand Strategy	Total Supplied (MCM)	Supply Increase (MCM)
New water sources (NWS)	1055	470
All existing and new buildings (RRWH_5)	880	295
Non-Revenue Water Reduction (NRWS)	837	252
All existing buildings (RRWH_1)	804	219
Selected existing and all new buildings (RRWH_4)	696	111
Only new buildings (RRWH_2)	661	76
Selected existing buildings (RRWH_3)	619	34
Efficient Water Use (EWU)	585	0
Reference	585	0

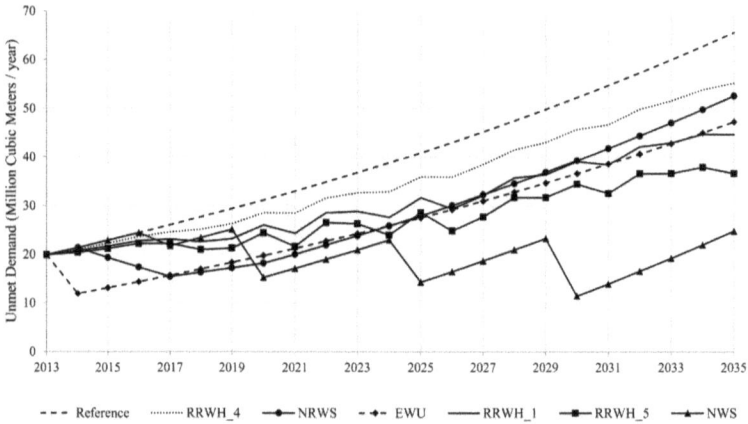

Figure 10. Unmet demand for Mombasa City for different management strategies (see Table 11 for acronym definitions).

For the combined scenarios, only the most feasible RRWH_4 is used, where only selected existing buildings and all new buildings implement RRWH. The results (Figure 11) show that before the year 2017, none of the combined strategies will meet the demand. Subsequently, only two scenario combinations, namely NWS/NRWS/EWU and NWS/RRWH_4/NRWS/EWU cover the demand fully from 2017 to the end of the simulation period. Consequently, meeting the demand before 2017 is not achievable by any of the investigated management strategies.

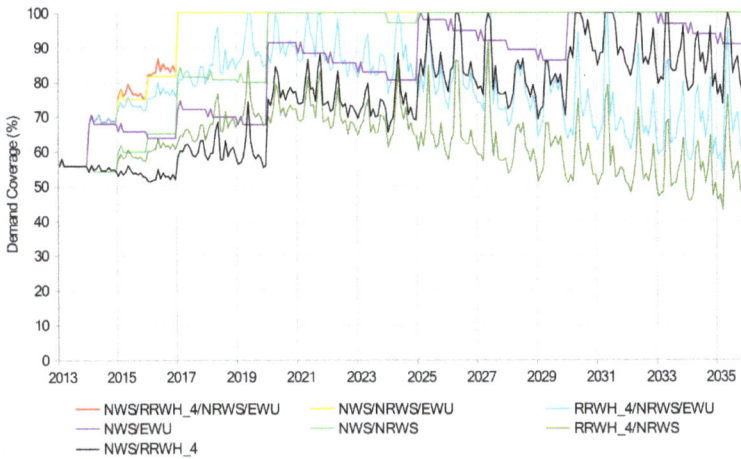

Figure 11. Demand coverage under different scenario combinations for Mombasa City.

The availability of water resources greatly depends on climatic conditions and the success of RRWH depends largely on the available precipitation. In this study, the reference scenario and other scenarios were based on the RCP 4.5 stabilization scenario with RCP 8.5 being used to understand how the system responds to a different climate change forcing scenario. The results show that the effect of predicted climate change considering RRWH_4 and RRWH_5 is not very appreciable (Figure 12). Considering RRWH_4, the average annual amount of rainwater harvested between 2014 and 2035 for RCP 4.5 and RCP 8.5 are 4.8 MCM and 4.7 MCM, respectively. The reduction of 2.5% is small compared to the uncertainty of the different GCM predictions. The two-sample t-test gives a p-value of 0.93, indicating no significant difference in the means at the 5% significance level. For RRWH_5, the average annual amount of rainwater delivered is 12.7 MCM for RCP 4.5 and 12.4 MCM for RCP 8.5, with a p-value of 0.94. Based on the results it can be concluded that the predicted effect of climate change on RRWH is negligible for Mombasa City.

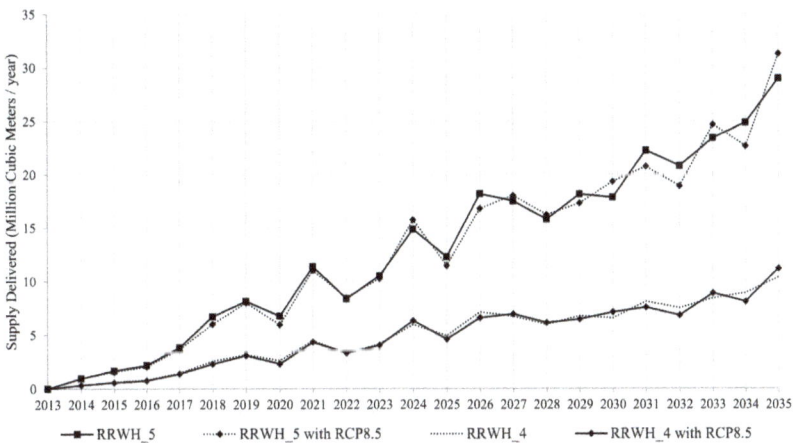

Figure 12. Effect of climate change on the RRWH_4 and RRWH_5 scenarios.

4. Conclusions

Overall, the results from this study show that the water demand for Mombasa City is expected to rise within 2014–2035 mainly due to socio-economic factors. Five possible RRWH implementation scenarios were established and investigated using a WEAP model. Using RCP 4.5 future climate data, the results showed that the average demand coverage improves significantly, mostly during the rainy season.

Comparing all of the management strategies, the development of new water sources (NWS) would lead to the highest demand coverage over the planning period up to 2035. None of the proposed strategies implemented in isolation will lead to full demand coverage. This means that NWS identified under the water supply master plan (2013) can only meet the demand if implemented alongside other strategies.

By combining the various strategies (only RRWH_4 used in the combinations) under a normal population growth rate, it was found that no combination of methods can cover the full demand for the entire 2014–2035 period. Under the highest water demand scenario, i.e., high population growth with better living standards, the city's water demand can never be met even under the most favorable climatic conditions. Any RRWH strategy is able to remedy but not completely solve the problem. The challenge to water managers is to make use of a combination of water supply sources, and to develop other strategies beyond those considered here e.g., river abstractions, desalination, or water reuse.

The results show that climate change will most likely not have an impact on the quantity of water delivered by RRWH systems in Mombasa. This means that RRWH can serve as a robust strategy against climate change effects.

The combination of image classification and water resource modelling proved to be a suitable tool for the development of roof rainwater harvesting strategies under changing water availability and demand. The efficiency of automatic image classification can be further improved by including height information obtained from stereo images.

This study mostly addressed the first part of RRWH supply systems at a larger scale and from a more technical perspective. Conveyance and storage systems have not been investigated in detail, and for the practical implementation of RRWH, socio-economic aspects and water quality issues should be considered as well. Additional studies are recommended on building conditions, roof rainwater quality, tank optimization/design, costs and fundraising, awareness creation, and sensitization of the city residents regarding RRWH. Another interesting area that could be investigated in the future is hydrological aspects, such as the benefit of RRWH for flood risk reduction and the effect of RRWH on groundwater recharge.

Acknowledgments: The work of Robert Ojwang was funded by Deutscher Akademischer Austauschdienst (DAAD) under the Entwicklungsbezogene Postgraduiertenstudiengänge (EPOS) program. The Digital Globe Foundation granted images. The Kenya Meteorological Department, Coast Water Services Board, and Mombasa Water and Sanitation Services Company provided data. We acknowledge the support of the aforementioned organizations. We thank the four anonymous reviewers for their comments, which helped to improve the manuscript. The publication of this article was funded by the Open Access fund of Leibniz Universität Hannover.

Author Contributions: Jörg Dietrich and Franz Rottensteiner developed the idea for the study. Matthias Beyer and Jörg Dietrich designed the model experiments; Robert Ojwang performed the model experiments; Robert Ojwang and Prajna Kasargodu Anebagilu analyzed the data; Robert Ojwang, Jörg Dietrich, Prajna Kasargodu Anebagilu, Matthias Beyer, and Franz Rottensteiner wrote the paper.

Conflicts of Interest: The authors declare no conflict of interest. The founding sponsors had no role in the design of the study; in the collection, analyses, or interpretation of data; in the writing of the manuscript, and in the decision to publish the results.

References

1. Abdulla, F.; Al-Shareef, A. Roof rainwater harvesting systems for household water supply in Jordan. *Desalination* **2009**, *243*, 195–207. [CrossRef]

2.   Siegert, K. *Introduction to Water Harvesting: Some Basic Principles for Planning, Design and Monitoring*; Water Reports FAO: Rome, Italy, 1994.

3.   Angrill, S.; Segura-Castillo, L.; Petit-Boix, A.; Rieradevall, J.; Gabarrell, X.; Josa, A. Environmental performance of rainwater harvesting strategies in Mediterranean buildings. *Int. J. Life Cycle Assess.* **2016**. [CrossRef]

4.   Gould, J.; Nissen-Petersen, E. *Rainwater Catchment Systems for Domestic Supply: Design, Construction and Implementation*; Intermediate Technology Publications: London, UK, 1999.

5.   Domènech, L.; Saurí, D. A comparative appraisal of the use of rainwater harvesting in single and multi-family buildings of the Metropolitan Area of Barcelona (Spain): Social experience, drinking water savings and economic costs. *J. Clean. Prod.* **2011**, *19*, 598–608. [CrossRef]

6.   Handia, L.; Tembo, J.M.; Mwiindwa, C. Potential of rainwater harvesting in urban Zambia. *Phys. Chem. Earth* **2003**, *28*, 893–896. [CrossRef]

7.   Thomas, T.H.; Martinson, D.B. *Roof Water Harvesting: A Handbook for Practitioners*; IRC International Water and Sanitation Centre: Delft, The Netherlands, 2007.

8.   Kahinda, J.M.; Taigbenu, A.E.; Boroto, R.J. Domestic rainwater harvesting as an adaptation measure to climate change in South Africa. *Phys. Chem. Earth* **2010**, *35*, 742–751. [CrossRef]

9.   Taffere, G.R.; Beyene, A.; Vuai, S.A.H.; Gasana, J.; Seleshi, Y. Reliability analysis of roof rainwater harvesting systems in a semi-arid region of sub-Saharan Africa: Case study of Mekelle, Ethiopia. *Hydrol. Sci. J.* **2016**, *61*, 1135–1140. [CrossRef]

10.   Lee, J.Y.; Bak, G.; Han, M. Quality of roof-harvested rainwater—Comparison of different roofing materials. *Environ. Pollut.* **2012**, *162*, 422–429. [CrossRef] [PubMed]

11.   Farreny, R.; Morales-Pinzo, T.; Guisasola, A.; Taya, C.; Rieradevall, J.; Gabarrell, X. Roof selection for rainwater harvesting: Quantity and quality assessments in Spain. *Water Res.* **2011**, *45*, 3245–3254. [CrossRef] [PubMed]

12.   Gikas, G.D.; Tsihrintzis, V.A. Assessment of water quality of first-flush roof runoff and harvested rainwater. *J. Hydrol.* **2012**, *466–467*, 115–126. [CrossRef]

13.   UNEP; CEHI. *A Handbook on Rainwater Harvesting in the Caribbean*; The United Nations Environment Programme (UNEP): Washington, DC, USA; The Caribbean Environmental Health Institute (CEHI): Castries, Saint Lucia, 2009.

14.   Ward, S.; Memon, F.A.; Butler, D. Performance of a large building rainwater harvesting system. *Water Res.* **2012**, *46*, 5127–5134. [CrossRef] [PubMed]

15.   Mehrabadi, M.H.R.; Saghafian, B.; Haghighi Fashi, F. Assessment of residential rainwater harvesting efficiency for meeting non-potable water demands in three climate conditions. *Resour. Conserv. Recycl.* **2013**, *73*, 86–93. [CrossRef]

16.   Meera, V.; Mansoor Ahammed, M. Water quality of rooftop rainwater harvesting systems: A review. *J. Water Supply Res. Technol. AQUA* **2006**, *55*, 257–268.

17.   Sazakli, E.; Alexopoulos, A.; Leotsinidis, M. Rainwater harvesting, quality assessment and utilization in Kefalonia Island, Greece. *Water Res.* **2007**, *41*, 2039–2047. [CrossRef] [PubMed]

18.   Mendez, C.B.; Klenzendorf, J.B.; Afshar, B.R.; Simmons, M.T.; Barrett, M.E.; Kinney, K.A.; Kirisits, M.J. The effect of roofing material on the quality of harvested rainwater. *Water Res.* **2011**, *45*, 2049–2059. [CrossRef] [PubMed]

19.   Byrne, J.; Taminiau, J.; Kurdgelashvili, L.; Kim, K.N. A review of the solar city concept and methods to assess rooftop solar electric potential, with an illustrative application to the city of Seoul. *Renew. Sustain. Energy Rev.* **2015**, *41*, 830–844. [CrossRef]

20.   Williams, N.; Quincey, D.; Stillwell, J. Automatic classification of roof objects from aerial imagery of informal settlements in Johannesburg. *Appl. Spat. Anal. Policy* **2015**, *9*, 269–281. [CrossRef]

21.   Veljanovski, T.U.; Kanjir, U.; Pehani, P.; Oštir, K.; Kovačič, P. Object-based image analysis of VHR satellite imagery for population estimation in informal settlement Kibera-Nairobi, Kenya. In *Remote Sensing-Applications*; InTech Europe: Rijeka, Croatia, 2012; pp. 407–434.

22.   Yates, D.; Sieber, J.; Purkey, D.; Huber-Lee, A. WEAP21—A demand-priority and preference-driven water planning model Part 1: Model characteristics. *Water Int.* **2005**, *30*, 487–500. [CrossRef]

23. Yates, D.; Purkey, D.; Sieber, J.; Huber-Lee, A.; Galbraith, H. WEAP21—A demand-priority and preference-driven water planning model. Part 2: Aiding freshwater ecosystem service evaluation. *Water Int.* **2005**, *30*, 501–512. [CrossRef]

24. Lévite, H.; Sally, H.; Cour, J. Testing water demand management scenarios in a water-stressed basin in South Africa: Application of the WEAP model. *Phys. Chem. Earth* **2003**, *28*, 779–786. [CrossRef]

25. Mutiga, J.K.; Mavengano, S.T.; Zhongbo, S.; Woldai, T.; Becht, R. Water allocation as a planning tool to minimize water use conflicts in the Upper Ewaso Ng'iro North Basin, Kenya. *Water Resour. Manag.* **2010**, *24*, 3939–3959. [CrossRef]

26. Falkenmark, M.; Lundquist, J.; Widstrand, C. Macro-scale Water Scarcity Requires Micro-scale Approaches: Aspects of Vulnerability in Semi-arid Development. *Nat. Resour. Forum* **1989**, *13*, 258–267. [CrossRef] [PubMed]

27. WRMA. *Integrated Water Resources Management and Water Efficiency Plan for Kenya*; Water Resources Management Authority (WRMA): Nairobi, Kenya, 2009.

28. Droogers, P.; Butterfield, R.; Dyszynski, J. *Climate Change and Hydropower, Impact and Adaptation Costs: Case Study Kenya*; Future Water: Wageningen, The Netherlands, 2009.

29. Taylor, K.E.; Stouffer, R.J.; Meehl, G.A. An overview of CMIP5 and the experiment design. *Bull. Am. Meteorol. Soc.* **2012**, *93*, 485–498. [CrossRef]

30. Overland, J.E.; Wang, M.; Bond, N.A.; Walsh, J.E.; Kattsov, V.M.; Chapman, W.L. Considerations in the selection of global climate models for regional climate projections: The Arctic as a case study. *J. Clim.* **2011**, *24*, 1583–1597. [CrossRef]

31. MWI. *Practice Manual for Water Supply Services in Kenya*; Ministry of Water and Irrigation (MWI): Nairobi, Kenya, 2005.

32. Mombasa County. *Mombasa County Government First County Integrated Development Plan (2013–2017)*; Government of Kenya: Mombasa, Kenya, 2014.

33. Kenya National Bureau of Statistics. *Kenya 2009 Population and Housing Census*; Government of Kenya: Mombasa, Kenya, 2009.

34. TAHAL/Bhundia Consultants. *Water Supply Master Plan for Mombasa and Other Towns within Coast Province*; Coast Water Services Board: Mombasa, Kenya, 2013.

35. WASREB. *A Performance Review of Kenya's Water Services Sector 2012–2013 (Impact Report Issue No. 7)*; Water Services Regulatory Board (WASREB): Nairobi, Kenya, 2014.

36. Lillesand, T.M.; Kiefer, R.W. *Remote Sensing and Image Interpretation*; John Wiley & Sons: New York, NY, USA, 1994.

37. CES Consultants. *Feasibility Study, Preliminary and Detailed Engineering Designs of Development of Mwache Multi-Purpose Dam Project along Mwache River—Hydrology Report*; Ministry of Regional Development: Nairobi, Kenya, 2013.

38. TWDB. *The Texas Manual on Rainwater Harvesting*; Texas Water Development Board (TWDB): Austin, TX, USA, 2005.

39. Stehman, S.V. Selecting and interpreting measures of thematic classification accuracy. *Remote Sens. Environ.* **1997**, *62*, 77–89. [CrossRef]

40. Grote, A.; Rottensteiner, F. Assessing the impact of digital surface models on road extraction in suburban areas by region-based road subgraph extraction. *Int. Arch. Photogramm. Remote Sens. and Spat. Inf. Sci.* **2009**, *38*, 27–34.

41. Hirschmüller, H. Stereo processing by semiglobal matching and mutual information. *IEEE Trans. Pattern Anal. Mach. Intell.* **2008**, *30*, 328–341. [CrossRef] [PubMed]

MDPI AG

St. Alban-Anlage 66

4052 Basel, Switzerland

Tel. +41 61 683 77 34

Fax +41 61 302 89 18

http://www.mdpi.com

Water Editorial Office

E-mail: water@mdpi.com

http://www.mdpi.com/journal/water